Selected Philosophical Essays

Carl Gustav Hempel (1905–1997) was one of the preeminent figures in the philosophical movement of logical empiricism. He was a member of both the Berlin and Vienna Circles, fled Germany in 1934, and finally settled in the United States, where he taught for forty-five years in New York, New Haven, Princeton, and Pittsburgh.

The essays in this collection come from the early and late periods of Hempel's career and chart his intellectual odyssey from a commitment to rigorous logical positivism in the 1930s (when Hempel allied himself closely with Carnap) to a more sociological approach that was close in spirit to the work of Neurath and Kuhn.

Most of these essays are hard to track down, and four of them are appearing in English for the first time. Cumulatively, they offer a fresh perspective on Hempel's intellectual development and on the origins and metamorphoses of logical empiricism.

Richard Jeffrey has prepared the collection for publication, and he has supplied introductory surveys to the essays as well as the brief biography on Hempel that begins on page viii.

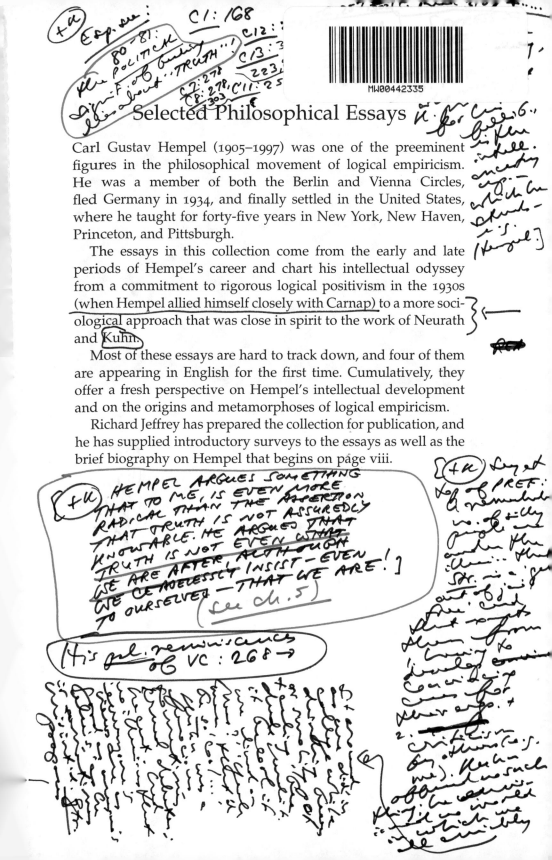

Richard Jeffrey (left) with the author, Carl G. Hempel (right).

Selected Philosophical Essays

CARL G. HEMPEL

Edited by RICHARD JEFFREY

CAMBRIDGE
UNIVERSITY PRESS

PUBLISHED BY THE PRESS SYNDICATE OF THE UNIVERSITY OF CAMBRIDGE
The Pitt Building, Trumpington Street, Cambridge, United Kingdom

CAMBRIDGE UNIVERSITY PRESS
The Edinburgh Building, Cambridge CB2 2RU, UK   http:www.cup.cam.ac.uk
40 West 20th Street, New York, NY 10011-4211, USA   http:www.cup.org
10 Stamford Road, Oakleigh, Melbourne 3166, Australia
Ruiz de Alarcón 13, 28014 Madrid, Spain

First published 2000

Printed in the United States of America

Typeface Palatino 10.5/13 pt.   *System* QuarkXPress [BTS]

A catalog record for this book is available from the British Library.

Library of Congress Cataloging in Publication Data

Hempel, Carl Gustav, 1905–
Selected philosophical essays / Carl G. Hempel; edited by Richard Jeffrey.
p.   cm.
Includes bibliographical references
ISBN 0-521-62448-7 – ISBN 0-521-62475-4 (pbk.)
1. Philosophy.   2. Empiricism.   I. Jeffrey, Richard C.   II. Title.
B945.H451   1999
191 – dc21

99–040313

ISBN   0 521 62448 7   hardback
ISBN   0 521 62475 4   paperback

Contents

Contents

Preface

Peter (= Carl Gustav) Hempel and I projected something like
this volume some six years ago as a representative selection of
his earliest and latest philosophical writings, a companion piece
to the familiar collection, *Aspects of Scientific Explanation*,[1] of
writings earlier and later than those. We began by translating his
1934 Berlin doctoral dissertation – a piece which I finally decided
to leave out. (It is available in the Hempel collection in the
Hillman Library of the University of Pittsburgh.) Included in its
stead are two articles (chapters 6 and 7) that are based on that
dissertation, which was published in the 1930s. The first one, a
translation from the German by Christoph Ehrlenkamp, has been
revised by Peter's old friend and collaborator, Olaf Helmer. The
only other paper we managed to translate was Peter's 1937 "Le
problème de la vérité," chapter 4 here. Other translators of essays
here were Wilfred Sellars (chapter 9) and Christian Piller (chapters 10 and 14).

In this book Peter speaks for himself; I have kept editorial
comment to a minimum. But I must give some account of his long
and remarkable life. The following brief biography is adapted
from *Erkenntnis* 47 (1997) 181–83, and it is reprinted here with
permission.

---

[1] Carl Gustav Hempel, *Aspects of Scientific Explanation* (New York: Free Press,
1965).

CARL GUSTAV HEMPEL
JANUARY 8, 1905–NOVEMBER 9, 1997

He lived in "interesting" times, which drove him out of Germany, first to Belgium and then to the United States – where he died sixty years later in Princeton, his adoptive home, much loved and full of honors.

Born in Oranienburg, near Berlin, he studied mathematics, physics, and philosophy at the universities of Göttingen, Heidelberg, Berlin, and Vienna, receiving his doctorate in Berlin just a week before Hitler assumed the mantle of Führer–Reichskanzler in 1934. Hempel had nearly finished his dissertation (*Beiträge zur logischen Analyse des Wahrscheinlichkeits-begriffs*) under the supervision of Hans Reichenbach when the latter was abruptly dismissed from his Berlin chair in 1933 shortly after Hitler become Chancellor. The ensuing problem of finding competent referees was sidestepped when Wolfgang Köhler and Nicolai Hartmann agreed to serve nominally in Reichenbach's place. By then Hempel and his wife, Eva Ahrends, had moved to Brussels, where his friend and collaborator Paul Oppenheim had made it possible for them to support themselves. In August 1937 they moved to Chicago, where Rudolf Carnap had obtained Rockefeller research fellowships for Hempel and his friend Olaf Helmer. From 1939 to 1940 he taught summer and evening courses at City College in New York before being appointed to an Instructorship at Queens College, New York; here he remained until 1948, having reached the rank of Assistant Professor. It was in this period that Eva Ahrends died, shortly after giving birth to their son. Three years later, Hempel married Diane Perlow. He was an Associate Professor at Yale University from 1948 to 1955, when he moved to Princeton as Stuart Professor of Philosophy, a post he held until mandatory retirement at age 68 in 1973. For the next two years he continued to teach at Princeton as a Lecturer, and then he moved to the University of Pittsburgh as University Professor of Philosophy in 1977. Upon his retirement from Pittsburgh in 1985, he returned to Princeton, where he continued his philosophical work for another decade. He is survived by his wife, Diane, and by his children, Peter Andrew Hempel and Miranda TobyAnne Hempel.

Although himself of unexceptionable "Aryan" stock, Hempel exhibited an insensitivity to such matters of a sort that constituted an offense in Germany at that time – namely, "Philosemitism," an offense against which his father and other well-wishers warned him more than once. Indeed, his wife Diane is Jewish; and his first wife, Eva, had inherited "Jewish blood" from her father, as had his teacher, Reichenbach, from his. That is what made Germany uninhabitable in 1934 for him as well as for them and made him slow to revisit the place after the war.

Jewish relns.

As a university student he had been preparing himself to teach mathematics in high school after receiving the doctorate. He thought he would have enjoyed that life. And at Princeton he generally chose to teach introductory courses – undergraduate courses in logic and the philosophy of science as well as graduate seminars in which he brought novices to the point at which they could make their own contributions. What lay behind that preference was not a logical empiricist's sense that more advanced students of vernacular philosophy would have been better off knowing less, but a kind of love or reverence or care for naïve minds, a sense, as he once put it, that they are the salt of the earth.

There was no arrogance in him; he got no thrill of pleasure from proving people wrong. His criticisms were always courteous, never triumphant. This quality was deeply rooted in his character. He was made so as to welcome opportunities for kindness, generosity, courtesy; and he gave his whole mind to such projects spontaneously, for pleasure, so that effort disappeared into zest. Diane was another such player. (Once, in a restaurant, someone remarked on their politeness to each other, and she said, "Ah, but you should see us when we are alone together. [Pause] Then we are *really* polite.") And play it was, too. He was notably playful and incapable of stuffiness.

Some have marvelled at his willingness to change his mind. Such philosophical lightness seems incompatible with the meticulous attention to details of arguments and definitions for which he is famous: the paradoxes of confirmation, the critique of the logical positivist conception of empirical meaningfulness, the analysis of explanation, and so on. But the lightness was a lack of nostalgia for bits of doctrinal baggage with their familiar stickers. He had no interest in ownership of such pieces; he really did just want to know the truth, and if a decade or two of close thought and argument should persuade him that some piece was empty, he would shove it overboard – with zest. He was rooted in truth no less than in kindness.

I do not expect to meet his like again.

Richard Jeffrey
Princeton, NJ
July 1998

Dates

CARL G. HEMPEL

Born in 1905 in Oranienburg, Germany. Studied mathematics, physics, and philosophy at the universities of Göttingen, Heidelberg, Berlin, and Vienna. Ph.D. University of Berlin, 1934.

Left Germany in 1934. Private research and writing in Brussels, Belgium, from April 1934 to August 1937. Research Associate in Philosophy, University of Chicago, 1937–38.

Instructor in philosophy in summer school, then in evening extension, at City College, New York, 1939–40.

Instructor, Assistant Professor of Philosophy, Queens College, Flushing, New York, 1940–48.

Associate Professor, Professor of Philosophy, Yale University, 1948–55.

Professor, Stuart Professor of Philosophy, Princeton University, 1955–73; Stuart Professor of Philosophy, Emeritus, 1973–97; Lecturer, 1973–75.

University Professor of Philosophy, University of Pittsburgh, 1977–85.

Visiting Professor at Columbia University, 1950; Harvard, 1953–54; Hebrew University, Jerusalem, March–July 1974; University of California, Berkeley, Winter Quarters, 1975 and 1977; Carleton College, 1976; University of Pittsburgh, 1976; University of California at Irvine, April–May 1978. Lectures in Beijing at

Peking University and before the Academy of Science; in Shanghai at Fudan University and before the Society of Díalectics of Nature.

Guggenheim Fellow, 1947–48; Fulbright Senior Research Fellow, Oxford University, 1959–60; Fellow, Center for Advanced Study in the Behavioral Sciences, 1963–64; Honorary Research Fellow in Philosophy, University College, London, 1971–72; Gavin David Young Lecturer, University of Adelaide, Australia, June–July, 1979; Fellow, Wissenshaftskolleg Zu Berlin, 1983–84; Fellow, The Mortimer and Raymond Sackler Institute of Advanced Studies, Tel Aviv University (February 15 to April 15, 1984).

[handwritten margin note: did you find his records there?]

President, American Philosophical Association, Eastern Division; Vice-President, Association for Symbolic Logic; Fellow, American Academy of Arts and Sciences; member, Académie Internationale de Philosophie des Sciences; member, American Philosophical Society; Corresponding Fellow, The British Academy; Fellow, Society of the Lincei (Rome).

Honorary doctorates:

Dr. of Science, Washington University, 1975
Dr. of Letters, Northwestern University, 1975
Dr. of Humane Letters, Princeton University, 1979
Dr. of Humane Letters, Carleton College, 1981
Dr. rer. pol., h.c., Freie Universität Berlin, 1984
D. Litt., University of St. Andrews, 1986
Ekrendoktorat der Philosophie, Johann Wolfgang Goethe-Universität, Frankfurt, 1987
Dr. of Letters, University of Pittsburgh, 1989
Doctor ad honorem in philosophy, University of Bologna, 1989
Honorary Doctorate, University of Konstanz, 1991

Died in Princeton, New Jersey, 1997.

Introduction

Philosophical Essays, Early and Late

The working title of this collection was "Philosophical Essays, Early and Late" – that is, earlier and later than the essays of Hempel's middle period, which are collected in his *Aspects of Scientific Explanation and Other Essays in the Philosophy of Science* [54].*
They were published in two decades (roughly, 1945–1965), during which logical empiricism evolved from an arcane neo-Kantian[1] or counter-Kantian sect into something that would seriously be called "The Received View"[2] – a view whose shape and reception owed much to those very essays. They were essays in the sort of logical constructivism that Russell had framed in *Our Knowledge of the External World as a Field for Scientific Method in Philosophy* (1915). Russell's book was the inspiration for Carnap's *Der logische Aufbau der Welt* (1928), and for the *Logische Syntax der Sprache* (1934), which incorporated Carnap's version of "physicalism." In those two decades Hempel could still regard the failings and limitations of that work as mere bumps on the road to logical empiricism. This book of Hempel's essays charts his course into and out of the methods and concerns of that middle period.

---

* Bracketed numbers refer to the list of Hempel's publications at the end of this volume.

[1] See Michael Friedman, "Overcoming Metaphysics: Carnap and Heidegger," in Ronald N. Giere and Alan W. Richardson, eds., *Origins of Logical Empiricism* (Minneapolis: University of Minnesota Press, 1996), pp. 45–79.

[2] See Frederick Suppe, ed., *The Structure of Scientific Theories*, 2nd ed. (Champaign: University of Illinois Press, 1977).

1

Having been one of the first logical empiricist graduate students,[3] Hempel would become the last survivor of the Berlin and Vienna circles. Throughout, he was in close intellectual contact with the major figures in the movement – and a movement it was, defined for its activists by no doctrine but by a certain morale and discipline. In his preface to the book that drew Hempel to Vienna, Carnap characterized that morale and discipline as follows[4]:

> Each collaborator contributes only what he can endorse and justify before the whole body of his co-workers. Thus stone will be carefully added to stone and a safe building will be erected at which each following generation can continue to work. . . . We feel all around us the same basic orientation, . . . which demands clarity everywhere, but which realizes that the fabric of life can never be fully comprehended. It makes us pay careful attention to detail and at the same time it makes us recognize the great lines which run through the whole. It is an orientation which acknowledges the bonds that tie people together, but at the same time it strives for free development of the individual. Our work is carried by the faith that this attitude will win the future.

Neurath put it differently; his famous nautical image compared the collaborators to "sailors who must rebuild their ship on the open sea, without ever returning to dry dock and constructing it anew out of the best materials."[5] For Neurath there were no logical building blocks; he saw clumpings (*Ballungen*) where Carnap would have clean-cut stones. In Vienna Hempel felt

---

[3] Herbert Feigl was earlier, with Moritz Schlick as *Doktorvater*. His dissertation was *Chance and Law* (Vienna University, 1927).

[4] Rudolf Carnap, *Der logische Aufbau der Welt* (Berlin-Schlachtensee: Weltkreis, 1928; Hamburg: Felix Meiner, 1961); Rolf A. George, trans., *The Logical Structure of the World and Pseudoproblems in Philosophy* (Berkeley: University of California Press, 1967).

[5] Nancy Cartwright et al., *Otto Neurath: Philosophy Between Science and Politics* (Cambridge, UK: Cambridge University Press, 1996), p. 89.

2

the pull of both tendencies, but it was Carnap's program of step-by-step construction, offering a chance for serious application of Russell's "Scientific method in philosophy," that inspired the work of his middle period. His later essays record a growing sense of that strategy as a dead end, at least for him – from which the way out might prove to be Neurath's *Gelehrtenbehavioristik*, empirical sociology of science, in something like its Kuhnian atavar.[6]

What was so attractive about the logical empiricist program (ca. 1934), when Hempel began writing these essays? In a word, the answer is *Principia*,[7] – that is, Whitehead and Russell's reduction of mathematics to logic, as seen through the lens of Carnap's 1934 reduction of logic to the syntax of the language of science. These "reductions" were problematical, but perhaps they were no more so than the recent reductions within physics of thermodynamics to statistical mechanics, gravitation to the geometry of space-time, or chemistry to quantum mechanics. The scientific philosophers of the Vienna Circle and the Berlin group hoped to resolve their problems in the scientific spirit just as Russell had urged in 1915 – that is, by careful analytical and constructive work.[8] The resulting aprioristic account of our knowledge of mathematics and logic would be compatible with a strict empiricism concerning factual matters, which lie outside mere syntax.[9] As to the detailed shape of syntax itself, that is ultimately a matter of choice for the scientific community, a matter in which

[6] See Thomas Kuhn, *The Structure of Scientific Revolutions* (Chicago: University of Chicago Press, 1962, 1970). Hempel's middle period writings and lectures provide many examples of his concern for historical and sociological detail in the philosophy of science (e.g., see his *Philosophy of Natural Science* [Englewood Cliffs, NJ: Prentice-Hall, 1966]).

[7] Bertrand Russell and Alfred Whitehead, *Principia Mathematica*, vol. 1 (Cambridge, UK: Cambridge University Press, 1910); 2nd edition (1925).

[8] See Michael Friedman, "Hempel and the Vienna Circle," in James Fetzer, ed., *Science, Explanation and Rationality: Aspects of the Philosophy of Carl G. Hempel* (Oxford, UK: Oxford University Press, forthcoming).

[9] For more about the program, see Ronald N. Giere and Alan W. Richardson, eds., *Origins of Logical Positivism* (Minneapolis: University of Minnesota Press, 1996).

3

the question of truth or falsity does not arise. What does arise in its place is the question of expediency for the purpose of saving the phenomena.[10]

And now it is time to turn to the "Truth" group of essays.

---

[10] Rudolf Carnap, *Der logische Syntax der Sprache* (Vienna: Julius Springer, 1934); Amethe Smeaton, Countess von Zeppelin, trans., *The Logical Syntax of Language* (London: Kegan Paul, 1937), sec. 82. The translated version was extensively revised by Olaf Helmer.

Truth

1 [2]* "On the Logical Positivists' Theory of Truth," *Analysis* 2 (1935), 49–59.
2 [6] "Some Remarks on 'Facts' and Propositions," *Analysis* 2 (1935), 93–6.
3 [7] "Some Remarks on Empiricism," *Analysis* 3 (1936), 33–40.
4 [12] "Le problème de la vérité," *Theoria* 3 (1937), 206–46. Translated here by the author as "The Problem of Truth." → /1937
5 [107] "The Signification of the Concept of Truth for the Critical Appraisal of Scientific Theories," *Nuova Civiltà delle Macchine* 8 (1990), 109–13. Here, "signification" is a misprint for "significance," which is itself a watering-down of "irrelevance" in the original typescript. This was reprinted under the title "The Significance of the Concept of Truth for the Critical Appraisal of Scientific Theories" (but it was listed in the table of contents as "Evidence and Truth in Scientific Inquiry") in William R. Shea and Antonio Spadafora, eds., *Interpreting the World* (Canton, MA: Science History Publications, 1992), pp. 121–9. Here, I follow the original typescript.

Hempel's defense of Neurath's and Carnap's physicalism in essays 1–3 testifies to the presence of certain "postmodern" themes in logical empiricism (= logical positivism):

A textualist turn to sentences from the facts or reality they are said to report.
A pragmatic turn from truth to apt inclusion in the text as the basic scientific concern.

QM
+ !
LE.

* The numbers in brackets represent the publication number for each chapter in this list. (See the section entitled "C. G. Hempel's Publications" beginning on p. 305.)

[HE DOWNPLAYS
ROLE OF
'TRUTH'
EARLY
ON]

A descriptive turn from logic to the empirical sociology of science.

Early (essays 1–4, in Brussels) and late (essay 5, in Princeton), Hempel saw <u>confirmation and acceptance</u>, not truth, as the crucial concept: The methodologically interesting question was how observation reports bring other sentences with them into the text of current knowledge – and how new entries bump old ones. For Hempel, the problem of analyzing the concept of confirmation of sentences by sentences is either the heart of the problem of truth or the successor to that problem. Thus, the early essay 4 <u>replaces the problem of truth</u> with the problem of analyzing the concept of confirmation of sentences by sentences, and the late essay 5 reiterates that theme in a Kuhnian mode.

[1937 !]

For the logical empiricists, *Principia Mathematica* was more than an attempt to reduce mathematics to logic. To that bare logical structure they would add empirical primitive symbols, and axioms involving those symbols that would make the system as a whole imply – in addition to the laws of pure mathematics – the laws of physics and, derivatively, <u>all of science.</u> This was the program of <u>physicalism</u> as Carnap saw it in 1934. It posited certain "protocol sentences" (*Protokollsätze*, which can be translated as "basic sentences" or "observation reports") to represent the empirical bases for knowledge. In the early 1930s a debate concerning the form and character of those sentences flared up in the pages of *Erkenntnis* and *Analysis*. <u>Neurath,</u> on the "left wing" of the Vienna Circle, saw protocol sentences – at least in principle – as the sort of thing that scientific observers could write on slips of paper and drop into a bin containing other observers' contributions, so that at any time the pooled contents could serve as the empirical basis of science at that time.[1] Schlick, on the "right," held that scheme to be unworkable in principle no less than in practice. He held that protocol sentences (his *Konstatierungen*) would need to be inseparably associated with observers and occasions in a way that would make them a useless

[i.e. qua sci. a priori] Köhler

---

[1] The slips of paper in the bin business is my way of telling Neurath's story, not his. (Editor)

jumble of unidentifiable context dependencies once they were mixed into the bin.

Essays 1–3 were young Hempel's contributions to the protocol-sentence debate. In essay 1 he describes it and defends the Neurath–Carnap left wing position. Essays 2 and 3 are further moves in the controversy, rejoinders to replies by Schlick and von Juhos.

In 1955 Hempel had still found it excruciating to recall those essays, which were then twenty years old – perhaps because of the scorn and ridicule Bertrand Russell had heaped upon them fifteen years earlier in Chapter X of *An Inquiry into Meaning and Truth* (1940), as in the following passage:

> ... To say: "A is an empirical fact" is, according to Neurath and Hempel, to say: "the proposition 'A occurs' is consistent with a certain body of already accepted propositions". In a different culture circle another body of propositions may be accepted; owing to this fact, Neurath is an exile. He remarks himself that practical life soon reduces the ambiguity, and that we are influenced by the opinions of neighbors. In other words, empirical truth can be determined by the police. ...[2]

Indeed, Hempel and Neurath had muddled the first of two distinctions that Carnap would propound at the Paris *Congrès International de Philosophie Scientifique* in 1935:

1. The question of the definition of *truth* must be clearly distinguished from the question of a criterion of *confirmation*.
2. In connection with confirmation two different operations have to be performed: the formulation of an observation and the confrontation of statements with each other; especially, we must not lose sight of the first operation.[3]

---

[2] Bertrand Russell, *An Inquiry into Meaning and Truth* (London: Geo. Allen and Unwin, Ltd., 1940), p. 148.

[3] Rudolf Carnap, "Wahrheit und Bewärung," *Actualités Scientifiques et Industrielles* 391 (Paris: Hermann & Cic, 1936), p. 23. The translated version of this essay, "Truth and Confirmation," by Feigl, appears in Herbert Feigl and Wilfrid Sellars, eds., *Readings in Philosophical Analysis* (New York: Appleton-Century-Crofts, 1949), pp. 126–7.

The second distinction did figure large in Neurath's and Hempel's view – but with a clarity limited by a muddling of the first distinction.

Essay 4 is a bridge both to Hempel's middle period and to essay 8 in the "Probability" group of essays. Here he still seems to be blurring Carnap's first distinction. (A clearer title for essay 4 might have been "The Problem of Confirmation"; and his contrast between absolute and apparent truths in the last paragraph might have been more clearly described as a contrast between sentences that are true and sentences that only seem true.) But the blur has a rhetorical use, for it matches the fog around the traditional philosophical problem of truth, which is the problem he means to dissolve here as a pseudoproblem.

Essay 5, which resonates strongly with contemporary work of Davidson, Rorty, and others,[4] can be read as a dramatic sermon on the text of Carnap's distinction 1 – and a startling corollary of the dissolution announced in essay 4.

[4] For references, see "Is Truth a Goal of Inquiry? Donald Davidson versus Crispin Wright" – that is, chapter 1 of Rorty's *Truth and Progress* (Cambridge, UK, and New York: Cambridge University Press, 1998).

Chapter 1

On the Logical Positivists' Theory of Truth[1]

This paper has been suggested by a recent discussion between Prof. Schlick and Dr. Neurath, made public in two articles which appeared in volume 4 of *Erkenntnis*,[2] which mainly concerns the positivistic concept of verification and truth.

For the following exposition it may be advantageous to refer to the well-known crude classification which divides the different theories of truth into two main groups – that is, the correspondence theories and the coherence theories of truth. According to the correspondence-theories truth consists of a certain agreement or correspondence between a statement and the so-called facts or reality; while according to the coherence theories truth is a possible property of a whole system of statements (i.e., a certain conformity of statements with each other). In extreme coherence theories truth is even identified with the mutual compatibility of the elements of such a system.

The logical positivists' theory of truth developed step-by-step from a correspondence theory into a restrained coherence theory.

---

[1] It has unfortunately been necessary to condense Dr. Hempel's paper slightly. (Editor [i.e., the editor of *Analysis*])

[2] M. Schlick, "Über das Fundament der Erkenntnis," *Erkenntnis* 4 (1934), 79–99. This essay was subsequently translated by Peter Heath as "On the Foundation of Knowledge," and it is reprinted in Henk L. Mulder and Barbara van de Velde-Schlick, eds., *Moritz Schlick: Philosophical Papers*, vol. 2 (Dortrecht: Reidel, 1979), pp. 370–88. See also O. Neurath, "Radikaler Physikalismus und 'wirkliche Welt,'" *Erkenntnis* 4 (1934), 346–62.

Let us shortly consider the most important logical phases of this process (which do not exactly correspond to the historical ones).

The philosophical ideas, which L. Wittgenstein has developed in his *Tractatus Logico-Philosophicus* and which represent the logical and historical starting point of the Vienna Circle's researches, are obviously characterized by a correspondence-theory of truth.

According to one of Wittgenstein's fundamental theses, a statement is to be called true if the fact or state of affairs expressed by it exists; otherwise the statement is to be called false. Now, in Wittgenstein's theory the facts constituting the world are conceived to consist ultimately of certain kinds of elementary facts which are not further reducible to other ones. They are called atomic facts, and those which are composed of them, molecular facts. In correspondence to these two kinds of facts, two kinds of statements are assumed: atomic statements that express the atomic facts, and molecular statements that express the molecular ones. The logical form in which a molecular statement is constituted by atomic ones reflects the formal structure of facts; and, consequently, just as the existence or nonexistence of a molecular fact is determined by the existence or nonexistence of its atomic constituents, the truth or falsity of a molecular statement is determined by the corresponding properties of the atomic statements. That is to say: each statement is conceived to be a truth-function of the atomic statements.

Wittgenstein's ideas concerning truth were rather generally adopted by the early Vienna Circle. The first to raise doubts, which soon developed into a very energetic opposition, was Dr. Neurath. The first who recognized the importance of Neurath's ideas was Prof. Carnap. He joined some of Neurath's most important theses and gave them a more precise form, and he and Neurath, mutually exciting each other, developed these ideas to the theory of truth which we shall deal with.

A crude, but typical, formulation of Dr. Neurath's main theses may be given as follows.[3]

---

[3] See Neurath, (1) "Soziologie im Physikalismus," *Erkenntnis* 2 (1931), 393–431; (2) "Physikalismus," *Scientia* 50 (1931), 297–303; (3) "Sozialbehaviorismus,"

Neurath
pushes
COHERENCE ;
theory vs. Witt.'s
CORRES-
PON-
DENCE
theory

Science is a system of statements which are of one kind. Each statement may be combined or compared with each other statement (e.g., in order to draw conclusions from the combined statements or to see if they are compatible with each other or not). But statements are never compared with a "reality," with "facts." None of those who support a cleavage between statements and reality is able to give a precise account of how a comparison between statements and facts may possibly be accomplished – nor how we may possibly ascertain the structure of facts. Therefore, that cleavage is nothing but the result of a redoubling metaphysics, and all the problems connected with it are mere pseudoproblems.

But how is truth to be characterized from such a standpoint? Obviously, Neurath's ideas imply a coherence theory.

Carnap developed, at first, a certain form of a suitable coherence theory, the basic idea of which may be elucidated by the following reflection: If it is possible to cut off the relation of sentences to "facts" from Wittgenstein's theory and to characterize a certain class of statements as true atomic statements, one might perhaps maintain Wittgenstein's important ideas concerning statements and their connections without further depending upon the fatal confrontation of statements and facts – and upon all the embarrassing consequences connected with it.

The desired class of propositions presented itself in the class of those statements which express the result of a pure immediate experience without any theoretical addition. They were called protocol statements, and they were originally thought to need no further proof.

Replacing the concept of atomic facts* by that of protocol statements was the *first* step in abandoning Wittgenstein's theory of truth.

A change of view concerning the formal structure of the system of scientific statements represents the *second* step of the

Sociologus 8 (1932), 281; (4) *Einheitswissenschaft und Psychologie* in the series *Einheitswissenschaft* (Vienna: Gerold, 1933); (5) "Protokollsätze," *Erkenntnis* 3 (1932), 204–14.
* Hempel's emendation. The *Analysis* version had "atomic statements."

evolution leading from Wittgenstein's theory of truth to that of Carnap and Neurath.

According to Wittgenstein, a proposition that cannot ultimately be verified has no meaning; in other words, a statement has a meaning when and only when it is a truth-function of the atomic propositions.

The so-called laws of nature, as will be illustrated below, cannot entirely be verified; therefore, they represent, according to the *Tractatus*, no statements at all, but they express mere instructions for establishing meaningful statements.

But when developing the theory I am speaking of, Carnap took into account both that in science empirical laws are formulated in the same language as other statements and that they are combined with singular statements in order to derive predictions. Therefore, he concluded that Wittgenstein's criterion for meaningful statements was too narrow and must be replaced by a wider one. He characterizes empirical laws as general implicative statements which differ by their form from the so-called singular statements, such as "Here is now a temperature of 20 degrees centigrade."

A general statement is tested by examining its singular consequences. But as each general statement determines an infinite class of singular consequences, it cannot be finally and entirely verified; it is only more or less supported by them. A general statement is not a truth-function of singular statements, but it has in relation to them the character of a *hypothesis*. The same fact may be expressed as follows: A general law cannot be formally deduced from a finite set of singular statements. Each finite set of statements admits an infinite series of hypotheses, each of which implies all the singular statements referred to. So, in establishing the system of science, there is a conventional moment; we have to choose between a large quantity of hypotheses, which are logically equally possible, and in general we choose one that is distinguished by formal simplicity as Poincaré and Duhem often accentuated.

Furthermore, it is important to recall that the singular statements have themselves the character of hypotheses in relation to the protocol statements, as Carnap shows in *Unity of*

Science.[4] Let us note that consequently even the singular statements which we adopt, which we regard as true, depend upon which of the formally possible systems we choose.

Our choice is logically arbitrary, but the large number of possibilities for choosing is practically restricted by psychological and sociological factors, as particularly Neurath emphasizes.

Thus also a second fundamental principle of the *Tractatus* must be abandoned. It is no longer possible to define the truth or falseness of each statement in terms of the truth or falseness of certain basic statements, whether or not* they may be atomic statements or protocol statements or other kinds of singular statements. For even the usual singular statements were revealed to be hypotheses in relation to the basic statements. Now a hypothesis cannot fully and finally be verified by a finite series of singular statements; a hypothesis is not a truth-function of singular statements, and consequently a singular statement which is not a basic statement itself is not a truth-function of basic statements.

So the refined analysis of the formal structure of the system of statements involves an essential loosening or softening of the concept of truth; for according to the considerations just mentioned, we may say: In science a statement is adopted as true if it is sufficiently supported by protocol statements.†

And that characterizes an essential trait which the theory considered here still has in common with Wittgenstein's view: the principle of reducing the test of each statement to a certain kind of comparison between the statement in question and a certain class of basic propositions which are conceived to be ultimate and not to admit of any doubt.

The *third* and *last* phase of the logical evolution here considered may be characterized as the process of eliminating even this characteristic from the theory of truth.

[4] Rudolf Carnap, *The Unity of Science* (London: Kegan Paul, 1934).

* Hempel's emendation. The words "whether or not" do not appear in the *Analysis* version.

† In the *Analysis* version this is followed by gibberish ("So there occurs in science, one drops at least one of the mentioned protocol sentences"), which should be compared with the end of the third paragraph below.

Indeed, as Dr. Neurath emphasized rather early, it is very well imaginable that the protocol of a certain observer contains two statements which contradict each other – say, "I see this patch entirely dark-blue and also entirely light-red." And if that occurs in science, one drops at least one of the mentioned protocol statements.

So protocol statements can no more be conceived as constituting an unalterable basis of the whole system of scientific statements, though it is true that we often go back just to protocol statements when a proposition is to be tested. Or, as Neurath says: We don't renounce a judge who decides if a statement in question is to be adopted or to be rejected; this judge is represented by the system of protocol statements. But our judge is removable. Carnap supports the same view, saying: There are no absolutely first statements for establishing science; for each statement of empirical character, even for protocol statements, further justification may be demanded (e.g., the protocol statements of a certain observer may be justified by statements contained in the report of a psychologist who examines the reliability of the observer before or even while he makes his observations).

So there may be attached to any empirical statement a chain of testing steps in which there is no absolute last link. It depends upon our decision when to break off the testing process, and it is no more exact to compare science with a pyramid rising on a solid basis. Neurath rather compares science with a ship which is perpetually being altered in the open sea and which can never be put into a dry dock and reconstructed from the keel upwards.

Obviously, these general ideas imply a coherence theory of truth. But it must be emphasized that by speaking of statements only, Carnap and Neurath do by no means intend to say, there are no facts, there are only propositions; on the contrary, the occurrence of certain statements in the protocol of an observer or in a scientific book is regarded as an empirical fact, and the propositions occur as empirical objects. What the authors do intend to say may be expressed more precisely thanks to

Carnap's distinction between the material and the formal mode of speech.[5]

As Carnap has shown, each nonmetaphysical consideration of philosophy belongs to the domain of logic of science, unless it concerns an empirical question and is proper to empirical science. And it is possible to formulate each statement of logic of science as an assertion concerning certain properties and relations of scientific propositions only. So also the concept of truth may be characterized in this formal mode of speech – namely, in a crude formulation, as a sufficient agreement between the system of acknowledged protocol statements and the logical consequences which may be deduced from the statement and other statements which are already adopted.

And it is not only possible but much more correct to employ this formal mode rather than the material one. For the latter involves many pseudoproblems which cannot be formulated in the correct formal mode.

Saying that empirical statements "express facts" – and consequently that truth consists in a certain correspondence between statements and the "facts" expressed by them – is a typical form of the material mode of speech.

The pseudoproblems connected with the material mode of speech are still alive in many an objection raised against Carnap's and Neurath's ideas; this is true also for some objections expounded in Prof. Schlick's paper (and for some considerations rather similar to them which von Juhos recently developed).[6]

[5] See Rudolf Carnap, *Der logische Syntax der Sprache* (Vienna: Julius Springer, 1934). The translated version of this book, which was extensively revised by Olaf Helmer, appeared in English as Countess von Amethe Smeaton Zeppelin, trans., *The Logical Syntax of Language* (London: Kegan Paul, 1937). See also Carnap, *Philosophy and Logical Syntax*, lectures given in London, 1934, reported in *Analysis* 2 (1934), 42–8 (and published in full as no. 70 in C. K. Ogden's "Psyche Miniature" monograph series [London: Kegan Paul, 1935], pp. 88–97); *The Unity of Science*, Psyche Miniatures 63 (1934). (This last is Max Black's translation of "Die physikalische Sprache als Universalsprache der Wissenschaft," *Erkenntnis* 2 [1931] 432–65 – ed.)

[6] B. v. Juhos, "*Kritische Bemerkungen zur Wissenschaftscheorie des Physikalismus,*" *Erkenntnis* 4 (1933), 397.

[: THE 'RELATIVISM' THREAT — IN '35.]

Prof. Schlick begins by raising the objection that radically abandoning the idea of a system of unalterable basic statements would finally deprive us of the idea of an absolute ground of knowledge and lead to a complete relativism concerning the problem of truth.

But we must reply that a syntactical theory of scientific verification cannot possibly give a theoretical account of something that does not exist in the system of scientific verification. And indeed, nowhere in science will one find a criterion of absolute unquestionable truth. In order to have a relatively high degree of certainty, one will go back to the protocol statements of reliable observers; but even they may give place to other well supported statements and general laws. So demanding an absolute truth criterion for empirical statements is inadequate; it starts from a false presupposition.

We may say that searching for a criterion of absolute truth represents one of the pseudoproblems that are due to the material mode of speech: indeed, the phrase that testing a statement is comparing it with facts will very easily evoke the imagination of one definite world with certain definite properties – and so one will easily be seduced to ask for the one system of statements which gives a complete and true description of this world and which would have to be designated absolutely true. By employing the formal mode of speech, we find that the misunderstanding which admits no correct formulation disappears – and with it the motive for searching for a criterion of absolute truth.

Prof. Schlick assumes an absolutely solid ground of knowledge; but on the other hand he concedes that it is advantageous in a theory of truth to consider only propositions. There is therefore only one way left for him to characterize truth – that is, to assume that there is a certain class of statements which are synthetic, and nevertheless absolutely, unquestionably true, by comparison with which every other statement might be tested. And in fact Prof. Schlick assumes that there are statements of this character; he calls them *Konstatierungen* ("statings"), and he attributes to them the form "Here now so and so" (e.g., "Here now blue and yellow side by side" or "Here now pain").

But Prof. Schlick himself concedes that any scientific statement

16

is a hypothesis and may be abandoned, and therefore he is obliged to assume that his unabandonable *Konstatierungen* are not scientific statements but that they represent the inducement for establishing protocol statements corresponding to them (e.g., "The observer Miller saw at that time and that place blue and yellow side by side").

Concerning these *Konstatierungen*, Prof. Schlick claims that (1) as distinct from ordinary empirical statements, they are understood and verified in *one* act (i.e., by comparing them with facts). So he returns to the material mode of speech, and he even describes *Konstatierungen* as the solid points of contact between knowledge and reality. The embarrassing consequences of such a form of consideration were pointed out just now. (2) Prof. Schlick assumes that *Konstatierungen* cannot be written down like ordinary statements and that they are valid only in one moment (i.e., when they are established). But then it is impossible to understand how a *Konstatierung* may be compared with an ordinary scientific statement. And such a comparison would be necessary, insofar as Prof. Schlick assumes that every empirical statement is in the end tested by *Konstatierungen*.

But it is important still to discuss the starting consideration of Prof. Schlick's ideas. This is the following one:

Carnap and Neurath's thesis, that in science a statement is adopted as true if it is sufficiently supported by protocol statements, leads into nonsense if the idea of absolutely true protocol statements is rejected; for obviously one may fancy many *different* systems of protocol statements and of hypothetical statements sufficiently supported by them; and according to Carnap and Neurath's formal criterion, each of these different systems, which may even be incompatible with each other, would be true. For any fairy tale there may be constructed a system of protocol statements by which it would be sufficiently supported; but we call the fairy tale false and the statements of empirical science true, though both comply with that formal criterion.

In short, what characteristics are there according to Carnap and Neurath's views, by which to distinguish the true protocol statements of our science from the false ones of a fairy tale?

As Carnap and Neurath emphasize, there is indeed no formal,

How distinguish [17] from 'fairy tale'?

no logical difference between the two compared systems, but there is an *empirical* one. The system of protocol statements, which we call true and to which we refer in everyday life and science, may only be characterized by the historical fact that it is the system which is actually adopted by mankind, and especially by the scientists of our culture circle; and the "true" statements in general may be characterized as those which are sufficiently supported by that system of actually adopted protocol statements.[7]

The adopted protocol statements are conceived as spoken or written physical objects, produced by the subjects just mentioned; and it might be possible that the protocol statements produced by different men would not admit the construction of one unique system of scientific statements (i.e., of a system sufficiently supported by the whole set of protocol statements of different people); but fortunately this possibility is not realized. In fact, by far the greater number of scientists will sooner or later come to an agreement, and so, as an empirical fact, a perpetually increasing and expanding system of coherent statements and theories results from their protocol statements.

In a reply to an objection of Zilsel,[8] Carnap[9] adds a remark which perhaps provides us a possibility of explaining that fortunate empirical fact.

How do we learn to produce "true" protocol statements? Obviously by being conditioned. Just as we accustom a child to spit out cherry-stones by giving it a good example or by grasping its mouth, we condition it also to produce, under certain circumstances, definite spoken or written utterances (e.g., to say, "I am hungry" or "This is a red ball").

And we may say that young scientists are conditioned in the same way if they are taught in their university courses to produce, under certain conditions, such utterances as "The pointer is now coinciding with scale-mark number 5" or "This

[7] So truth is not reduced without qualification to formal properties of a system of statements: Carnap and Neurath do not support a pure coherence-theory – but, as we expressed it at the beginning, a restrained coherence-theory of truth.

[8] Zilsel, "*Bemerkungen zur Wissenschaftslogik*," *Erkenntnis* 3 (1932), 143.

[9] Carnap, "*Erwiderung auf Zilsel und Duncker*," *Erkenntnis*, 3 (1932), 177.

word is Old-High-German" or "This historical document dates from the 17th Century."

Perhaps the fact of the general and rather congruous conditioning of scientists may explain to a certain degree the fact of a unique system of science.

The evolution of the concept of truth we considered is intimately allied to a change of view concerning the logical function of protocol statements. Let me finish with some remarks relating to this point.

Originally, Carnap introduced the concept of protocol statements to denote the basis of testing empirical statements; in divorce from Wittgenstein's principles, he showed that even singular statements have the character of a hypothesis in relation to the protocol statements: They cannot be finally verified, but they may only be more or less confirmed by them. And there is no precise rule stipulating a minimum degree of confirmation as necessary for a statement to be adopted. In the end, the adoption or the rejection of a statement depends upon a decision.

"NO PRECISE RULE"

And in the recent form of Carnap and Neurath's theory, protocol statements are still more radically divested of their basic character: They lose the irremovability originally attributed to them. Even the protocol statements are revealed to be hypotheses in relation to other statements of the whole system; and so a protocol statement, like every other statement, is at the end adopted or rejected by a decision.

So I think that there is no essential difference left between protocol statements and other statements.

Dr. Neurath proposes to confine the term *protocol statements* to statements of a certain form (i.e., to those in which the name of an observer and the result of an observation occurs). Hereby he will accentuate the empirical character of science, in which a thorough testing is mostly traced back to observation statements.

Prof. Carnap on the other hand accentuates that (1) a test is not in all cases traced back to such observation statements and that (2) observation-statements of the kind Dr. Neurath means may themselves be tested by a reduction to statements, even of a different form. And (3) he emphasizes that in any case determining the formal characteristics of protocol statements is a matter of

convention, not a question of fact. He illustrates this point of view
by sketching three different conventions, each of which may
be chosen in order to characterize formally a class of protocol
statements. One of these conventions has been suggested by Dr.
Popper; it consists of admitting statements of any form to figure
as protocol statements. Prof. Carnap finds Dr. Popper's convention to be the most suitable and most simple of the three
conventions discussed by him. And indeed I think that just this
convention suits in the most perfect manner Carnap and
Neurath's general views of verification and truth.

Thus, the concept of protocol statements may have become
superfluous at the end. But it was at least a most essential auxiliary concept, and relativizing or fully abandoning it represents
the last step in an extended theoretical development.

Finally, we consider the consequences this evolution has for
the problem of atomic facts, which plays a considerable part in
Wittgenstein's theory.

Thanks to correctly expressing the problems to be solved in
the formal mode of speech, the double question of what are
atomic facts – and what are atomic statements – is revealed to be
only one question, at first expressed in the material and then in
the formal mode.

So there remained only one problem – i.e., to find the structure of the atomic statements – or in the version due to Prof.
Carnap, to find the logical form of protocol statements. This
problem was at first (e.g., in *Unity of Science*) conceived to be a
question of fact; but then, Carnap's considerations led to the
result that the form of protocol statements cannot be found but
must be fixed by a convention. And this insight eliminates from
the Logical Positivist's theory of verification and truth a remain-
der of absolutism which is due to metaphysical tendencies and
which cannot be justified by a correct syntactical analysis of
science.

Chapter 2

Some Remarks on "Facts" and Propositions

(1935)

1. In a recent article in *Analysis* (vol. 2, no. 5), Prof. Schlick traces the outlines of his view concerning the relationship of propositions to "facts." In this account, Prof. Schlick makes a contribution for which we must be grateful by elucidating some essential points of his article "Das Fundament der Erkenntnis" (*Erkenntnis* 4 [1933], 79), which occasioned a logical controversy, the fundamental ideas of which I tried to characterize in my note "On the Logical Positivists' Theory of Truth" (*Analysis*, vol. 2, no. 4).

In his paper, Prof. Schlick raises certain objections to some of the considerations which I sketched in my article and which correspond to Dr. Neurath and Prof. Carnap's view; he gives his objections the form of questions, which may be summed up by asking: What harm is there in saying that propositions are compared with "facts" and that true propositions express "facts"? To this question I shall try to reply.

2. Prof. Schlick illustrates the character of the comparison between a proposition and "facts" by means of a very instructive example. But I think that in an essential respect his account is not quite adequate.

For on the one hand Prof. Schlick expressly describes propositions as empirical objects (of a special kind) which may be compared with any other empirical object. So far I fully agree with him.[1] But if we take him at his word, we must expect that the

---

[1] Prof. Schlick's new explanations (pp. 66–7) show that concerning the character of propositions, the difference between his and Dr. Neurath's view is not

21

proposition he chooses as an example will be tested by comparing the *physical object* (consisting of ink-symbols) "This cathedral has two spires" (or a similarly shaped physical object which is to be found in Prof. Schlick's Baedeker) with another physical object called the cathedral. Such a comparison may very well be realized (it would lead to such statements as: The proposition contains more parts, called *"words,"* than the cathedral has spires); but evidently it does not permit us to test the proposition (indeed, there is no specific "correspondence" between the two compared physical objects). But here Prof. Schlick introduces a second interpretation of the comparison, saying that "it is done by looking at the cathedral and at the sentence in the book and by stating that the symbol 'two' is used in connection with the symbol 'spires,' and that I arrive at the same symbol when I apply the rules of counting to the towers of the cathedral" (*Analysis*, vol. 2, no. 5, p. 67). Here, he evidently compares the proposition in his Baedeker with *the result* (not with the act!) *of his counting* the spires; this result may have the form "I now see two spires" or something like that, but in any case it is a second *proposition* with which the first is compared. (And now there really is a certain congruence, because in this case both propositions contain the word "two.")

Thus, Prof. Schlick's example reveals that speaking of a "comparison between a proposition and 'facts'" is nothing but an abbreviated and convenient method of describing a comparison between certain propositions (and just this is meant by saying that in "a logical respect" "propositions cannot be compared to anything but propositions"). Such a comparison

nearly so great as it might seem from Prof. Schlick's article in *Erkenntnis*; his recent formulations reveal almost a verbal congruence with certain statements of Dr. Neurath's in *Erkenntnis* 4 (1934), especially pp. 355–6, which Dr. Neurath thought to be in contradiction to Prof. Schlick's view – and likewise with my remarks in *Analysis* 2, 4 (1935), pp. 54 and 57.

It might have been better to employ the term *sentence* to designate the series of symbols we are speaking of; for the English word *proposition* is also used with other meanings. But as the latter term has been employed in the two articles in *Analysis* which we refer to, I have refrained from replacing it by the univocal word *sentence*.

refers to the logical relations which hold between the compared propositions.

In order to find out, for example, if a certain hypothesis h is confirmed or falsified by the "observed facts" with which it is compared, one has to ascertain if the observation statements are *compatible* with (or even *deducible* from) h – or if they *contradict h*. Such a logical (syntactical) examination of propositions may be performed, as Carnap has shown in his *Logische Syntax der Sprache*[2] (cf. also "Philosophy and Logical Syntax," London, 1934[3]), without knowing the meaning of the propositions, by a mere comparison of the symbols which the propositions are composed of. (Stating that both the propositions mentioned above contain a sign shaped "two" is an example of this kind of comparison.)

3. But furthermore, Carnap has shown that the logical relations which hold between two propositions depend upon the syntactical rules of the language which we choose. A proposition p may be deducible from a proposition h with respect to one system of rules – and not deducible from h with respect to another one. Therefore, the result of what is called a "comparison between propositions and 'facts'" depends upon the syntax of scientific language – a circumstance which need not necessarily, but will at least very easily, be veiled by the material mode of speech, the latter evoking in the imagination the belief that the "facts" with which propositions are to be confronted are substantial entities and do not depend upon the scientist's choice of syntax-rules.

4. This point is also fundamentally connected with the question of the "structure of facts." If one admits this expression, it seems to be legitimate, for example, to ask if the structure of facts

---

[2] Rudolf Carnap, *Der Logische Syntax der Sprache* (Vienna: Julius Springer, 1934). The translated version of this book, which was extensively revised by Olaf Helmer, appeared in English as Countess Amethe Smeaton von Zeppelin, trans., *The Logical Syntax of Language* (London: Kegan Paul, 1937).

[3] This paper, a lecture, was given in London in 1934, and it was reported in *Analysis* 2, 3 (1934), pp. 42–8. It was subsequently published in full as no. 70 in C. K. Ogden's "Psyche Miniature" monograph series (London: Kegan Paul, 1935), pp. 88–97.

admits only the occurrence of rational values of the different coefficients of physical state, or if, on the contrary, the old principle *natura non facit saltus* is valid in the sense that also irrational values are physically possible. And if one thinks, further, that the structure of propositions must be in a certain way isomorphous to the structure of facts, one has to ask: Have we to introduce the system of real numbers in order to give a true image of the structure of facts? But both these questions are pseudoproblems; for it is impossible to imagine an experience which might furnish a decision by falsifying one of the two possibilities. It is a question of syntactical convention whether to admit or to exclude the occurrence of irrational-number-symbols – and thereby to stipulate a rational or irrational metrical structure for "facts."

5. In Carnap and Neurath's theory of science, the empirical character of scientific research is fully maintained. It is expressed by emphasizing that scientific propositions are tested by such statements (often observation statements) as have been produced or adopted by instructed observers or "scientists." (Giving someone "deictic definitions" – of the kind Prof. Schlick speaks of – is a special way of instructing or "conditioning" him for the production of observation statements.)

But I think it is not quite harmless (though, of course, not "false" either) to say that those observation statements (and the statements supported by them) "express 'facts'"; for this term indicates something which is once and forever fixed with all its characteristics, whilst it is essential for the system of scientific statements that it may always be changed again, that no proposition is adopted once and for all, and in addition that the adoption of any observation statement has, after all, the character of a *convention*. But even if one denies this, as Prof. Schlick does, it is not harmless to say that "propositions express 'facts.'" For one system of observation statements (or other basic statements) is compatible with many different systems of physical statements (see page 12), so that any of the ordinary physical statements, even statements such as "This is a piece of iron," is a hypothesis, the adoption of which depends in the end upon a convention.

And the character of statements which are adopted by a convention evidently does not admit of such questions as: Are there statements which express facts adequately, which are absolutely true (are perhaps the *Konstatierungen* of this kind)? Prof. Schlick had dealt with questions of this type, and I therefore put forward in the paper which has been mentioned certain scruples concerning the formulation that "propositions express 'facts.'"[4]

---

[4] Some months ago, Dr. Popper published a most suggestive book, *Logik der Forschung* (Vienna: Springer, 1935), in which he examines, amongst other questions, some of the problems touched in the present discussion and gives a detailed account of the conventional character of the basic propositions in science.

Chapter 3

Some Remarks on Empiricism

(1936)

In this note, I want to deal with the objections which Dr. v. Juhos has raised in this periodical[1] to certain considerations concerning the language of science, which are mainly due to Carnap, Neurath and Popper, and some which I have outlined in an earlier paper.[2]

We may distinguish three main points to which v. Juhos's criticism refers: (I) The "behavioristic" interpretation of psychological statements such as "I see blue": (II) the view that in our empirical science even statements of the kind just mentioned might be "altered" or "abandoned" under certain conditions; (III) the proposal to express "epistemological" considerations in the formal mode of speech.

Von Juhos maintains that each of these points is a thesis or a postulate of physicalism: He speaks of "the physicalist mode of speech which demands the alterability of all statements" (loc. cit., p. 89; in this passage, elements of each of the three independent points are combined), and he declares the recommendation of the formal mode to be "a special postulate of Physicalism" (loc. cit., p. 90). For the sake of clarity, it ought to be noticed that only (I) is a thesis of physicalism, whilst (II) and (III) are fully independent of it.

---

[1] B. v. Juhos, "Empiricism and Physicalism," *Analysis* 2 (1935), 81.
[2] Carl Hempel, (1) "On the Logical Positivists' Theory of Truth," *Analysis* 2 (1935), 49. (See Chapter 1 here.)

Indeed, physicalism[3] asserts that the language of physics is a universal language of science – that is, "every sentence of any branch of scientific language is equipollent to some sentence of the physical language and can therefore be translated into the physical language without changing its content."[4] This thesis obviously entails the special assertion that the statements of psychology can be translated into the physical language.[5] In the Vienna Circle, this was originally called the *thesis of behaviorism.* At present, this designation is rarely employed, in order to avoid confusing this logical thesis, which deals with the syntax of scientific language, with the psychological statements and the methodological principles of American behaviorism. But as v. Juhos employs the term "behaviorism" in order to denote the syntactical thesis under consideration, it may be suitable to do the same in the following discussion.

So (I) is, in fact, a thesis of physicalism. But (III) is independent of physicalism: The thesis of behaviorism as well as many

---

[3] See, for example, Carnap, (1) *The Unity of Science* (London: Kegan Paul, 1934); (2) *Philosophy and Logical Syntax* (London: Kegan Paul, 1935), pp. 88–97; (3) *Testability and Meaning,* which was published in *Philosophy of Science* 3 (1936), 419–71, and 4 (1937), 1–40; (4) "Über die Einheitssprache der Wissenschaft," which was published as part of a series of monographs entitled *Actualités Scientifiques et Industrielles* 389 (Paris: Hermann & Cie, 1936), pp. 60–70. See also Neurath, (1') "Physicalisme," *Scientia* 80 (1931), 117; (1") "Physikalismus," *Scientia* 50 (1931), 297; (2) "Physicalism," *The Monist* (1931), 618–23; (3) *Empirische Soziologie* (Vienna: Julius Springer, 1931); (4) "Soziologie im Physikalismus," *Erkenntnis* 2 (1931), 393. See also Moritz Schlick, (1) "De la relation entre les notions psychologiques et les notions physiques," *Revue de Synthèse* 10, (1935), 5–26.

[4] Carnap (2), p. 89. In the meantime, Carnap has developed a certain modification of physicalism (see his papers [3] and [4]). For our purposes it is, however, not necessary to refer to this refined form of the physicalistic thesis.

[5] For a detailed account of this thesis, see, for example, Carnap, (5) "Psychologie in physikalischer Sprache," *Erkenntnis* 3 (1932), 107; (6) "Les concepts psychologiques et les concepts physiques sont-ils foncièrement différents?" *Revue de Synthèse* 10 (1935), 43–53. See also Carl Hempel, (2) "Analyse logique de la psycholgie," *Revue de Synthèse* 10 (1935), 27–42 (Chapter 9 of this book), where much additional literature is indicated. See also Neurath, (5) *Einheitswissenschaft und Psychologie* (Vienna: Gerold, 1935).

other theses of this kind can be formulated both in the material[6] and in the formal mode. And obviously point (II) is not a consequence of physicalism either; this is illustrated by the facts that in earlier papers serving to establish the thesis of physicalism (especially in [1]), Carnap started from the assumption that science is based on protocol statements which need no confirmation (and are, therefore, not alterable); and that Schlick in a recent paper[7]) expressly acknowledges physicalism though rejecting the assertion (II).

On (I).   Let us now consider first those of v. Juhos's objections which really concern physicalism, namely, those directed against (syntactical) behaviorism.

(*a*) In a rough formulation,[8] the thesis of behaviorism (physicalism) asserts that for any statement speaking of "feelings," "thoughts," "acts of will," etc., of a person, there is an equipollent statement which speaks exclusively of the "bodily" behavior (movements, sounds pronounced, physiological reactions, etc.) of the person in question; and v. Juhos is right in asserting that, according to behaviorism, the indications of a man about the state of his own feelings (e.g., his saying or writing, "I feel pain") have the same logical function for testing a psychological hypothesis ("This person is in the pain-state") as the pointed, written, or spoken indications of a barometer have for testing a physical hypothesis ("In the interior of this instrument, the air-pressure is now 1 atmosphere").

As a consequence, incorrect indications of the human "pain-indicator" correspond, for behaviorism, to incorrect indications of an air-pressure-indicator or some other instrument. But saying, as v. Juhos does, that for this reason "the physicalist . . . must in cases of the above-mentioned type reproach the instruments in question with a lie or an error" (loc. cit., p. 87) is evidently running just contrary to the fundamental idea of behaviorism:

---

[6] See, for example, Carnap, (1) "Erwiderung auf die . . . Aufsätze von E. Zilsel und K. Duncker," *Erkenntnis* 3 (1932), 177–88 (especially pp. 183–7, where he points out the dangers which arise from employing the material mode in the discussion of the syntactical thesis of behaviorism).

[7] Schlick (1), p. 14.   [8] See Hempel (2).

That attitude would mean applying to physics the point of view of (a rather primitive form of everyday-life) psychology, whilst behaviorism tends to proceed vice versa.

In fact – and this is accentuated by behaviorism – the methodological principles of scientific psychology are the same as those of physics. And particularly in the case of incorrect indications, one blames neither the instrument nor the test-person; one rather tries to *explain* the occurring "deviations" (e.g., by showing that certain disturbing factors are working). Formally, such an explanation consists of the establishment of a hypothesis concerning disturbing factors.

(*b*) In psychology, two particularly important kinds of hypotheses of this type are error-hypotheses and lie-hypotheses.

Now, v. Juhos thinks (loc. cit., p. 84, ff.) that to explain incorrect indications of a man about his own state of feeling, physicalism demands (or at least admits) the introduction of an error-hypothesis; and he asserts that this "has no meaning for the Empiricist" (loc. cit., pp. 85–86). It ought to be noticed, however, that *physicalism makes not the least allegation concerning the question of lie and error*, as may be seen for the above formulations and from the papers cited. Therefore, even if v. Juhos were right in asserting that scientific psychology does not admit of error-hypotheses in the considered cases, this would not have consequences for physicalism.

Besides that, it seems to me that his assertion is not right:

Allowing, or generally forbidding, the introduction of error-hypotheses concerning the results of psychological self-observation ("introspection") is, I think, a question of syntactical convention, which cannot be true or false, but only more or less practical. In the case of allowing, the convention mainly consists of specifying how to test such an error-hypotheses. Hereby, the error-hypotheses are *given* a "meaning."

As to actual scientific psychology, I think that it would *not* generally exclude error-hypotheses concerning the results of self-observation. For example: Any psychologist would admit the possibility (and even the frequent occurrence) of error when a man states through self-observation the motives of his actions – for example, when he says, "I change my political party out of conviction; I am sure that I have not got the least desire for material advantages." (In cases of this type, psychoanalysis, e.g., furnishes us with certain testing-methods.) And it seems to me that there is only a difference of degree between this case and that of a person saying "I feel pain" or "I see blue."[9]

And a more general point of view: If lie-hypotheses were not admitted in the latter case, a sentence of the form "When pronouncing the statement *S*, the person *A* committed an error" would be meaningful or meaningless (not false!) – that is, it would belong or not belong to scientific language – depending on the words occuring in *S*; and this would be so even if *S* itself ("I feel pain") were a formally correct and meaningful statement of the language under consideration. This would obviously be very little expedient. Therefore, I think that scientific psychology would not exclude once and for all error-hypotheses concerning statements like "I see blue."

However, this question cannot be discussed here in more detail, because it is independent of the problems of physicalism.

(*c*) Von Juhos further objects to physicalism leading to such "absurd" "consequences" (loc. cit., p. 85) as the assertion that people who do not know the pain-state-criteria of contemporary psychophysiology do not mean anything when speaking about pains. This *is* absurd, indeed, but it *is not* a consequence of physicalism. Physicalism refers only to the fact that all the methods employed in everyday life and by science for testing statements about people's feelings, thoughts, and the like consist of examining their behavior. In everyday life, one refers principally to such characteristics of behavior as a play of features, gestures,

---

[9] v. Juhos says that "while formulating a proposition such as 'I feel pain,' I know already whether I have made a true or a false statement" (loc. cit., p. 84). I think that with equal justice, one could say, "When formulating a statement about the real motives of my action, I know already if I am right or wrong."

In quite a similar connection, v. Juhos asserts that, concerning the falseness of contradictory statements, an error is impossible and that somebody who pretends that a certain contradiction is true must purposely make a false statement (loc. cit., pp. 89–90). From the experience I acquired when teaching mathematics at school, I think that on this point v. Juhos is certainly not right.

blushing, crying, etc. Scientific psychology, which has at its disposal many more empirical laws than everyday life, knows that those characteristics are empirically connected with other ones (e.g., certain nervous reactions), which, consequently, may likewise serve for testing the statement in question. But the introduction of new testing-methods does not render the older ones "meaningless" (though they may turn out to be less exact) – just as in physics the introduction of a new (e.g., electrical) method of testing a temperature-indication does not imply the rejection of the former methods (such as using a mercury-thermometer or even judging by simply touching the body in question) as "meaningless."

(*d*) Finally, v. Juhos gives an example (loc. cit., pp. 88–9) meant to illustrate the possibility of testing certain psychological statements directly – that is, without referring to the "bodily" behavior of the person in question. Without discussing this general question here (it had better be put into the formal mode before being examined), I want to remark that the example given by v. Juhos seems to me not to stand its ground. For how can the blind man (B) first know that his colour-experiences occur in regular connection with those of the seeing man (S) and that they may, therefore, serve him in the future for "directly" controlling S's color-statements? At first, S must tell him one or several times when he, S, is seeing blue. And B, when noticing that each time he himself has likewise a color-experience, may state a general law: Each time I have a color-experience, S has one, too. And this law may serve him for further controlling S. But B would *not* be able to establish the law *without* knowing first the indications of S, which are at the same time *a part of S's bodily behavior.*

Concerning (II). (*e*) It first has to be noticed that the conception of certain statements as unalterable or the stipulation that no statement of empirical science has to be maintained at all costs is a matter of convention between scientists; this has been accentuated particularly by Carnap.[10]

---

[10] See Carnap, (8) "Über Protokollsätze," *Erkenntnis* 3 (1932), 215. See also Carnap (3).

On the other hand, one may ask, which of the two possible conventions is in better congruence with the actual methodological attitude of our empirical science? As to this question, first Neurath[11] and Popper,[12] later joined by Carnap in their view, have insisted upon the alterability of *any* statement in empirical science; but they do *not* pretend that "we are allowed to alter the propositions obtained by observation as it pleases us" (v. Juhos, loc. cit., p. 83). On the contrary, a statement which is once acknowledged (and in particular an observation-statement) can be abandoned only if it is incompatible with another statement which is confirmed by observation in a very high degree.[13]

(*f*) As to the self-observation statements, there exists at least the possibility of introducing a lie-hypothesis, and therefore it is possible that even a statement of this kind should be abandoned. Dr. v. Juhos himself admits, "In stating 'I feel pain' I may have said something wrong . . . because I wanted to lie" (loc. cit., p. 84). And this evidently invalidates his assertion that statements such as "I feel pain" "are unalterable, for the possibility of error concerning them does not exist" (p. 84). No: Such a statement might be altered – at least by adopting a lie-hypothesis. As to the question of this being the only possibility, see (*b*).

Moreover, I wish to emphasize that when discussing the possibilities of lie and error in this connection, v. Juhos seems to start from a misunderstanding similar to that concerning physicalism (see [*b*]): The adherents of the alterability of any scientific statement do *not* pretend that the alteration of the considered psy-

---

[11] See, for example, Neurath, (6) "Protollsätze," *Erkenntnis* 3 (1932), 204; (7) "Radikaler Physikalismus und 'wirkliche Welt,'" *Erkenntnis* 4 (1934), 346; (8) "Pseudorationalismus der Falsifikation," *Erkenntnis* 5 (1935), 353. (Here in *Pseudorationalismus*, Neurath expressly accentuates that observation statements – for which he proposes a certain common form – possess a "greater stability" [p. 362] than many other statements in empirical science: They need not be altered so often as the latter ones.)

[12] See K. Popper, (1) *Logik der Forschung* (Vienna: Julius Springer, 1935). This book contains a detailed theory of the principles of testing in science and in particular an important and original account of the alterability of any empirical statement.

[13] For more detail, see particularly Popper (1).

chological statements must always be carried out by assuming an error; the distinction between lie and error is quite independent of the problem in question.

On (III). (*g*) Finally, v. Juhos develops a very interesting objection against the proposal of employing the formal mode of speech[14] in epistemological discussion (loc. cit., pp. 90–2). He refers to a formulation given by me: "The system of protocol statements which we call true . . . may only be characterized by the historical fact, that it is . . . actually adopted . . . by the scientists of our culture circle." Dr. v. Juhos translates this into the formal mode as follows: "[T]he system of protocol statements that we call true is characterized by the quality that certain protocol statements belong to it, which assert that this very system of statements is acknowledged as true by the scientists of our cultural circle" (loc. cit., p. 91). From this remark, v. Juhos deduces the "objection . . . that an infinite number of systems of protocol statements which are not to be contradicted might be quoted, all of which contain those particular statements, which characterize as true the system of protocol statements of our science, but for the rest are incompatible with this system. And all these imaginable systems have to be considered true exactly in the same sense as our true science" (loc. cit., p. 91).

However, the translation given by v. Juhos is not adequate. The term *historical fact* serves to express a reference to that which is *acknowledged* as factual *by our science*. For the translation into the formal mode, we have to replace "It is a fact that . . ." with "The statement . . . is sufficiently confirmed by the protocol statements adopted in our science." So, the following translation of the cited passage results: "The following statement is sufficiently confirmed by the protocol statements adopted *in our science*: 'Amongst the numerous imaginable consistent sets of protocol statements, there is in practice exactly one which is adopted by the vast majority of instructed scientific observers; at the same time, it is just this set which we generally call true.'" The whole

---

[14] See, for example, Carnap (1), p. 37 ff.; (2), p. 68 ff. See also Carnap, (9) "On the Character of Philosophic Problems," *Philosophy of Science* I (1934), 5–19, 251.

is by no means a contradictory proposition, nor can one deduce from it the consequences indicated by v. Juhos.

In the end, v. Juhos's objection furnishes us with a new example in favor of the proposal to use the formal mode of speech in logical discussion; for, as we saw, my earlier material formulation caused a misunderstanding which is now, I hope, cleared up.

Chapter 4
The Problem of Truth[1]

(1937)

I. INTRODUCTION

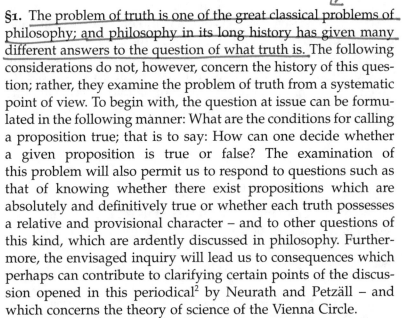

§1. The problem of truth is one of the great classical problems of philosophy; and philosophy in its long history has given many different answers to the question of what truth is. The following considerations do not, however, concern the history of this question; rather, they examine the problem of truth from a systematic point of view. To begin with, the question at issue can be formulated in the following manner: What are the conditions for calling a proposition true; that is to say: How can one decide whether a given proposition is true or false? The examination of this problem will also permit us to respond to questions such as that of knowing whether there exist propositions which are absolutely and definitively true or whether each truth possesses a relative and provisional character – and to other questions of this kind, which are ardently discussed in philosophy. Furthermore, the envisaged inquiry will lead us to consequences which perhaps can contribute to clarifying certain points of the discussion opened in this periodical[2] by Neurath and Petzäll – and which concerns the theory of science of the Vienna Circle.

The method by which we will attempt to establish the criteria

---

[1] *Change title.* (Italicized footnotes were added by Hempel, in German, shortly after publication. – Editor.)

[2] Compare Neurath (8); Petzäll (2). – Numbers in parentheses refer to the bibliography at the end of this article.

for truth will not be that of synthetic and speculative philosophy; it will rather consist in a logical and methodological analysis of the procedures which are employed, in science, to verify a proposition; for it is only by such an analysis that one can hope to determine the character of truth in the only sense that matters for our scientific and everyday knowledge.

This analytic manner of treating philosophical problems has been developed mainly by the Vienna Circle and some other thinkers with a similar attitude, such as Reichenbach and Popper. To express it in the manner of Moritz Schlick, the founder and spiritual father of the Vienna Circle, this analytic method (which, according to him, is simply *the* philosophical method) serves to determine the *meaning* of the problem in queston.[3]

For our analysis of verification we will divide all sentences into two classes: first, the class of the empirical propositions, which describe, as one says, the facts of the world of our experience and the laws that govern them; and second,[4] the class of the propositions of logic and mathematics. As we shall see, the verification procedures and, consequently, also the characteristics of the so-called "truth" of the propositions of these two classes are quite different; for this reason we will examine them separately, in sections II and III.

II. VERIFICATION AND "TRUTH" IN THE EMPIRICAL SCIENCES[5]

§2. Before we enter into our analysis, let us recall briefly the close relations which exist between the concept of *empirical truth* and that of *empirical reality*. In fact, it is customary, particularly in philosophy, to say both that the empirically true propositions repre-

---

[3] For the application of this point of view to the problem of truth, compare also the elementary and very instructive brochure of Hans Hahn ([1a], [1b]).

[4] *Here a language governed exclusively by L-transformation rules is presupposed. In a more general case there could also be among the symbolic sentences P-determinate ones; these are then true! (Lutman)*

[5] The considerations of this section are based essentially on the writings of Carnap (whose recent publication [7] presents the deepest logical analysis of the problem), and of Neurath and Popper.

sent or describe empirical reality and that a proposition is true if, in this reality, the fact affirmed by the proposition exists. One has even frequently attempted to define the concept of *empirical* truth with the help of this formula. But such a definition is worthless, since the concept of reality it uses is still more problematic than that of truth; therefore, we shall attempt on the contrary to clarify its meaning with the help of the concept of truth. An analysis of the concept of empirical truth will at the same time be an analysis of the concept of empirical reality, for to say that such and such a phenomenon is empirically real is nothing else than to say that the proposition describing the phenomenon is empirically true. This observation brings us back to our problem of knowing the verification procedures for empirical propositions.

§3. Let us distinguish between direct and indirect verification. *Direct verification* is based on simple observations, on immediate perceptions, and it applies to such propositions as "There is a lamp on this table." *Indirect verification* applies above all to propositions which cannot be verified by immediate observations, as in the case of the kinetic theory of heat. To be sure, we cannot draw a neat boundary between these two forms of verification; it is rather a matter of difference in degree, and there are an infinity of intermediaries between cases of quite direct verification and cases of an extremely mediated and indirect verification. But, in our context, a vague distinction like the one we have just illustrated by extreme cases turns out to be sufficient.

§4. *Let us first consider indirect verification*, for example, as applied to the proposition "The base of this lamp is iron." In order to examine the validity of this proposition, one can refer to the magnetic properties of iron, or, more precisely, to the empirical law that says that a magnet and an iron object exert a mutual attraction on each other. One can, therefore, verify the proposition by bringing a magnetized needle close to the base of the lamp and seeing whether it turns toward the base. Evidently, there are still many other means of verifying the proposition (e.g., one can measure the density of the substance in question or can observe the optical spectrum of its vapor, etc.).

Now, all these different procedures of indirect verification have the same logical structure: They consist of deducing from the proposition to be verified consequences which admit of direct verification – and which one verifies by immediate observations. In fact, in our example, we have deduced from the proposition "The base of this lamp is iron" a consequence which admits of direct verification, namely, "The magnetized needle turns toward the base of the lamp."

Let us examine two more examples. One of the most beautiful methods of verification for the hypothesis that the molecules of a substance – say, of air – are in a perpetual state of agitation rests on the observation of the Brownian motion. The reasoning underlying this method of verification can be summarized in the following manner: Air molecules cannot be made visible in a microscope; however, if in the air one suspends particles which are very light but are still visible in the microscope, one will observe that these particles move irregularly because of the impacts on them of the molecules. Hence, if the molecules of air are in thermal motion, then particles of tobacco, suspended in the air, will perform an irregular dance which one can observe in the microscope; that is the Brownian motion. Then here, too, one deduces from a proposition which is not directly verifiable a consequence capable of verification by direct observations.

Finally, let us take an example of verification as actually done in daily life. This example will show us at the same time that there are no differences of principle between the methods of scientific verification and those used in ordinary life. One day, about to go out, I find that I have no gloves. I remember that I still had them with me yesterday morning and that then took a taxi. I suppose that it is there that I forgot them. How to verify this hypothesis? A direct observation is evidently not possible, so I reason as follows: If the driver is honest, he will have returned the gloves to the taxi company, which customarily sends the objects forgotten in its cars to the police Lost and Found Department. If I go there to ask for my gloves, I will then be able to verify by direct observation whether or not my hypothesis was true.

Perhaps it will be objected that this latter case differs from the others in that in this deduction, there are still other premises than the proposition that is to be verified. In fact, in order to arrive at

(margin note: THE LOST GLOVES)

our directly verifiable consequence, we have introduced the hypothesis that the driver was honest – and still others – and we must presuppose that these accessory propositions are true in order to arrive at our conclusion. But it is exactly the same in every other case of verification. Thus, in our first example, the proposition "A magnetized needle will undergo a deviation" cannot be deduced from the hypothesis "The base of the lamp is iron" as its only premise. Still others are needed (e.g., notably, the law of magnetic attraction, for evidently it is only thanks to this law that one can arrive at the indicated conclusion). Furthermore, this deduction relies on the following premise: Near the base of the lamp there is a magnetized needle; for the law of attraction alone would not allow us to deduce from our hypothesis the consequence that a magnetic needle will undergo a deviation.

In sum, our antecedent indications must be made precise in the following manner: To verify in an indirect manner an empirical proposition – let's call it P – one adds to P certain other propositions P_1, P_2, P_3, ... (i.e., in general, propositions which have already been verified and which one adopts as valid). From this class of premises one deduces, by logical transformations, a consequence C which admits of direct verification. In our example, P would be the proposition "The base of this lamp is iron," P_1 would be the law of magnetic attraction, P_2 the proposition "Near the base of the lamp there is a magnetized needle," C would be the proposition "The needle will undergo a deviation in the direction of the base of the lamp."

By analogous considerations, one easily applies this logical schema to the verification of the hypothesis of the lost gloves, and one sees, also, that empirical verification possesses the same logical character in science as in daily life.

Moreover, we can now understand something more important from the logical point of view: The logical structure of the verification of an individual proposition is the same as that of a universal proposition. We understand here[6] by *individual proposi-*

---

[6] To avoid recourse to logistic, we will limit ourselves here to illustrative remarks; although these do not represent a precise definition, they will be sufficient for what follows.

tion a proposition which refers to a particular object or event localized in a certain finite domain of space-time (e.g., the proposition "This object is iron" or "Yesterday, I lost my gloves in the taxi"). By contrast, we call a *universal proposition* every assertion concerning all objects or all events of a certain kind, whatever may be their spatio-temporal location. This form of proposition is of particular importance in science. Indeed, all "empirical laws" are universal propositions. Thus, for example, the hypothesis of molecular movement asserts that for any substance, whatever be its location in the universe, its molecules are at each instant in movement, with velocities corresponding in a well-determined manner to the temperature of the substance. This hypothesis is thus a universal proposition. As we have just seen, the verification of this hypothesis consists of deducing empirical consequences which admit direct verification (e.g., "In the microscope, one will see, under certain conditions, little points in irregular movement"). To arrive at this consequence, one uses, as in the other cases, not only the hypothesis to be verified but also accessory premisses – premisses (which, by the way, are very complicated in this case) comprising among other things the theory of the microscope. The logical schema underlying the verification of this universal proposition is, thus, the same as that which we have found for individual propositions.

Let us now return after this remark to our general analysis, which evidently is not yet completed with the establishment of this schema. *Let us grant that the consequence C, which one has theoretically deduced from the proposition P, be confirmed by direct observation.* Can one then say that P is true? Certainly not. For, to illustrate it by our first example, a magnetic needle which one brings close to the base of the lamp also undergoes a deviation if the base of the lamp is made not of iron but of nickel. C can therefore be confirmed even if P is false.

The fact that the confirmation of C does not suffice to prove the truth of P appears also in a very clear fashion for universal propositions. In fact, an observation of the Brownian motion of particles of smoke suspended in air certainly does not suffice to make true the general proposition that the molecules of each substance are in thermal agitation. But even if one limits oneself

No *possibility* *of "complete" i.e. verification of truth*

to verifying the hypothesis of molecular movement for air, then one or even a hundred or a million observations of Brownian movement in the air would not suffice to prove the truth of the hypothesis; for the latter affirms that at any moment in any place where there is air, this molecular movement occurs. This universal proposition comprises, as one can see, an infinity of individual propositions which in principle cannot possibly all be verified, since one can never perform more than a finite number of observations. A universal proposition, therefore, does not admit complete verification – nor, therefore, a complete and definitive proof of its truth.

A similar situation holds for individual propositions. In fact, each individual proposition admits an infinity of consequences. The assertion of the truth of an individual proposition, therefore, brings with it the same assertion for an infinity of other propositions, for which it will again be impossible to speak of complete verification. The assertion, for example, that the base of this lamp is made of iron implies, by the law of magnetic attraction, the assertion that any magnet undergoes at each instant an attraction in the direction of the base of the lamp (i.e., an infinite number of consequences); and in taking account of the chemical, optical, and other properties of iron, one can deduce innumerable further consequences; and again one cannot verify them all. *It is, therefore, impossible in the verification of an empirical proposition*, whether individual or universal, *ever to arrive at a complete and definitive proof of its truth.*

No final truth

Let us examine on the other hand the possibility of the proof of the falsity of an empirical proposition. Let us assume that *a directly verifiable consequence of a proposition P is not confirmed by an observation*: In this case, does *P* have to be regarded as false? At first glance, it does seem that already the nonconfirmation of a single consequence *C* is sufficient to infirm, *modo tollendo*, the proposition *P*. Let us admit, for example, that in the case of the hypothesis that a certain object is iron, a magnetic needle which one brings close to it does not deviate. Does one not have to conclude that the hypothesis is false; and, generally speaking, is it not a sufficient condition for us to know that *P* is false that one of the consequences *C* of a proposition *P* is not confirmed? It is not so; that can be seen by the following consideration: What we have just called nonconfirmation by observation of the consequence C consists of the establishment, under the circumstances described by the premises, of an *observation proposition O* which contradicts C. If, for example, one has deduced from a hypothesis the consequence that under such-and-such circumstances the needle of a measuring instrument must lie between lines 5 and 6 on the scale, one would speak of the nonconfirmation of this consequence if an observer were to find that under the indicated conditions the needle lies near line 9; this finding, which constitutes an observation sentence O, contradicts C.[7] On the other hand, as we have seen, C is not a logical consequence of the proposition P alone, but of an entire class of premises $P, P_1, P_2, P_3, \ldots$. In short, in the case of nonconfirmation of a hypothesis P, the logical situation is therefore the following: The premises $P, P_1, P_2, P_3, \ldots$ imply C, and the observation proposition O contradicts C (and, consequently, the premises $P, P_1, P_2, P_3, \ldots$); in other words, the propositions $P, P_1, P_2, P_3, \ldots$, and O are incompatible; to avoid a contradiction, at least one must be eliminated. But it is not necessarily the proposition P which one must give up; and in fact, in such cases, empirical science sometimes eliminates P, sometimes one of the other premises, and sometimes O. Let us consider examples.

The first of the three possibilities is that which, at first glance, seems to be the only imaginable one: since the observation, formulated in the proposition O, contradicts C, one gives up the proposition P as infirmed. This method will in general be applied in cases like that of the magnetic examination of a substance which one assumes to be iron. If the observation does not confirm the prediction that a magnetic needle brought near it undergoes a deviation, then one will give up the hypothesis that the substance is iron.

But even in this case it is possible that one should maintain P

---

[7] The case of confirmation of a hypothesis by an observation (or, as one often says, by the facts) can be characterized in the same sense, and it is in this sense that Neurath says, in his discussion remark in *Theoria*, that scientific verification of a proposition consists of comparing this proposition with others which one has already verified.

despite the nonconfirmation of C. In fact, it is possible that an examination will lead us to abandon one of the premisses P_1, P_2, P_3, ... It can turn out, for example, that the needle used has lost its magnetism or that it is stuck. It is true that in general one uses, aside from P, only premisses which have been verified and which are relatively well confirmed, so that one is not easily disposed to abandon one of them; but since, as we have seen before, no proposition can ever be confirmed completely and definitively, it remains always possible to discard, upon further examination, one of the premisses which until then were rather well confirmed – and to maintain the proposition P. This eventuality is even of particular importance: Some of the greatest advances in empirical science have consisted of abandoning, in such cases of nonconfirmation as we are studying, a premise which had been considered as unshakeable, and of replacing it by a new and unexpected hypothesis. Such was the case, notably, in the development of the theory of relativity. This theory is based, among other things, on the abandonment of the classical concept of simultaneity, which figured among the premisses considered until then as absolutely indubitable and even logically evident.

It remains for us to consider the third possibility: that of abandoning O in order to avoid a contradiction. One will perhaps be disposed to think that at least this case cannot occur in science, for O expresses only what one has directly observed; and one will say that what the observer has seen, he has seen. One cannot deny or reject his observation proposition. And yet, even this possibility is not finally excluded in science. There are two principal cases where this possibility is realized: first, that of an observational error. If, for example, all the measurements which an observer has made to verify an empirical law square with it very well, except for a single one, one will assume, in general, that the observer has made a mistake in observing his measuring instrument or in recording the observed value. That is to say, one gives up the observation proposition which expresses this measurement and one eliminates it from the system of propositions considered sufficiently confirmed. Secondly, it happens sometimes that an observer knowingly falsifies his observational results or

1. [margin]

2. [margin]

[margin note right] EVEN OBSERVATIONAL ERRORS DATA IS NOT TOTALLY TRUSTWORTHY

[EPIG. for Element 118 anecdote]

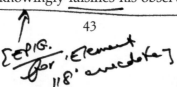

that he invents them without having carried out the necessary observations. It is evident that, when such a case is known, one gives up as at least dubious all the observational propositions which this investigator has ever established.

So we find, in sum, that if one of the directly verifiable consequences of a proposition P is infirmed by observation, it is not always the proposition P which is abandoned; it is always possible that an empirical examination leads science rather to renounce one of the other propositions which play a role in the verification – and that the proposition P be retained. An empirical proposition thus no more admits a complete and definitive proof of falsehood than of truth: *Each empirical proposition is only a hypothesis* in the sense that *it is capable neither of being proved nor disproved in a complete and definitive fashion;*[8] the observations forming the foundation of all empirical verification can only *confirm to a certain degree* the proposition which they serve to test. If, for example, the test, by direct observation, of different consequences C_1, C_2, C_3, ... of a proposition P yields positive results without any consequences of P being known to be infirmed by observation, one will say that the degree of empirical confirmation of P increases along with the series of favorable results.

Yet, whatever may be the number of favorable cases: it is logically impossible ever to arrive at a complete proof of the truth of P. For this reason it is preferable not to speak of empirical truth or falsity. In fact, what one usually understands by truth and falsity is, so to speak, of a ungraduated, definitive character, which does not permit saying that one proposition is more true than another, nor that such and such a proposition is perhaps true today but will not be true tomorrow. On the contrary, the notion of empirical confirmation which we have introduced in order to take account of the structure of all empirical verification possesses a provisional and graduated character: It is provided with an index of degree and an index of time. Any empirical

---

[8] I cannot therefore be in agreement with the *Falsifizierbarkeitsprinzip* of Popper. Compare on this subject Popper (1), Neurath (6), Carnap (7), Grelling (1), Hempel (3).

hypothesis can be confirmed only to a certain degree, which is determined by the experiences (formulated in the observation propositions) which one has available at the time of the verification.

As for the concept "degree of confirmation of a hypothesis with respect to a given class of reference propositions," we have limited ourselves here to illustrating its significance without, however, establishing an exact definition. It seems, furthermore, to be rather difficult to determine this concept in a metric fashion. Perhaps one could define it in a purely topological fashion; one would then limit oneself to arranging the hypotheses, without ascribing numerical degres of confirmation to them, in a schema which would not even be linear – and in which the better confirmed of two hypotheses would precede the less well confirmed. A precise definition of what we have called the degree of confirmation would then have to consist of establishing general criteria which make it possible to arrange the empirical hypotheses in such a scheme.[9]

§5.[10] The question "Can't the degree of confirmation of a hypothesis be metricized?" is, by the way, still under lively discussion. It is above all Reichenbach who gives an affirmative answer to this question; namely, he interprets the degree of confirmation of a hypothesis as a probability.

Now, it is true that in science, just as in ordinary language, one often says that a hypothesis is more or less probable – when one wants to indicate that it is more or less well confirmed by experience. But this manner of speaking refers in general not to a well determined concept of probability; the importance of Reichenbach's thesis consists precisely in the fact that it asserts this formulation and makes it precise by resting on a considerably deepened theory of probability.[11]

---

[9] For more details concerning the definition of such topological concepts, see Hempel and Oppenheim (1).

[10] Readers not interested in the relations between the concepts *degree of confirmation* and *probability* may omit this paragraph.

[11] Reichenbach develops this theory in a detailed fashion in his book (1). As for his conception of the probability of a hypothesis, he has given supplementary explications in his article (2).

According to this theory, which rests on the empirical-statistical interpretation of probability, every probability statement (*Wahrscheinlichkeisaussage*) establishes a relation between a certain infinite series of events $(x_1, x_2, x_3, \ldots)$ and two characteristics f and g; namely, it asserts that between the x_i that have the characteristic f, the fraction of those which also possess the characteristic g converges toward a number p which is called the probability with which f brings g with it in the series of events under consideration. Reichenbach represents this statement by a formula of the following form:

(EP) $\qquad (i)\{f(x_i) \xrightarrow[p]{} g(x_i)\} \quad i = 1, 2, 3, \ldots$

(i.e., for every i, if x_i has the property f, then, with probability p, x_i also has the property g).[12]

Let us assume, for example, that the x_i series is an indefinitely extendable series of rolls, performed with a well determined die, that f is the property of being an honest roll, that g is the property of giving 2 as a result, and that $p = 0.17$. Then the statement in question asserts that the probability of rolling a 2 with this die is equal to 0.17; and that means that the percentage or "the relative frequency" of rolls which yield a 2 converges toward 0.17.

Now the question that interests us here is whether this concept of probability – which usually is called "probablity of events" – can also be applied to hypotheses and empirical theories. (There is no difference in principle between theories and hypotheses, as every theory can be considered to be a logical product of hypotheses[13] – and, thus, a single complex hypothesis.)

Reichenbach gives an affirmative answer to this question. He bases this answer on the consideration that a theory or universal hypothesis can be considered as a very extended set of affirmations, each one of which refers to a particular case; thus, for example, quantum theory, which he chooses as an example, allows us to make certain forecasts which have the form of indi-

---

[12] In the publications cited, Reichenbach generally introduces two series of events (x_i) and (y_i); however, it is always possible to put his probability statements in the form (EP), where only one series occurs.

[13] One obtains the logical product of two proposition by connecting them with "and."

vidual propositions, such as "At such and such a moment, the Geiger counter in my laboratory will click." Reichenbach also makes the assumption that a theory can be regarded as a logical product of individual propositions, and he defines the probability of the theory by the relative frequency of those among these propositions that are confirmed by the experiment.[14]

Now, I think that this definition of the probability of a theory, although seemingly very plausible and adequate, encounters rather grave difficulties.

1. One of these difficulties has already been highlighted by Popper (in his book [1]), where he discusses in a systematic and critical manner several methods which one could try for defining the probability of a hypothesis, and where he arrives at the conclusion that the empirical-statistical concept of probability which we have illustrated above cannot be applied to a hypothesis or theory. Among other things,[15] Popper has, namely, called attention to the fact that a theory cannot be considered as an assemblage of individual assertions. In effect, the universal hypotheses of empirical science have a conditional form; in the simplest case such a hypothesis can be represented as follows (Reichenbach adds a further symbol of probability, as we shall soon see):

$$(H) \qquad (x)\{F(x) \supset G(x)\}$$

That is to say: "For all places x in the world: if $F(x)$, then $G(x)$" – for example, "If, at no matter which place x in the world, there is iron at 1200°C, there is light radiation at that place x." Now, such "universal implication" permits us, it is true, to assert individual propositions (having the form, "$G(a)$") – but only on the condition that another proposition ("$F(a)$"), which, itself, is not a consequence of the theory, has already been established as empirically confirmed. A *theory* alone, without such individual propo-

---

[14] See Reichenbach (2), §IV. It is this concept – called by Reichenbach "probability of the first form" – which we will limit ourselves to in our examination here. In (2), Reichenbach introduces, furthermore, the concept "probability of the second form," whose definition, however, is based on that of the first form and is therefore equally touched by the following considerations.

[15] Popper (1), pp. 191–2.

sitions which serve as accessory premisses, *affirms, therefore, no individual proposition of the kind envisaged by Reichenbach, nor can it be interpreted as a logical product of such propositions.* But it is just on this interpretation that Reichenbach bases his application of the concept of probability to theories.

2. In another paragraph of his accont,[16] Reichenbach presents in a formal manner the interpretation of hypotheses which forms the basis of his definition of "probability of a hypothesis." He states, namely, that each hypothesis can be considered to be a statement of the form (EP); one would, therefore, have to replace the representation (H) of a hypothesis with the following formula:

$$(HP) \qquad (i)\{F(x_i) \xrightarrow[p]{} G(x_i)\} \quad i = 1, 2, 3, \dots$$

where p is a more or less high degree of probability. And from this representation, Reichenbach concludes equally that a proposition can be considered as a denumerable series of individual propositions.

Now, this formal representation of hypotheses highlights a second difficulty encountered by Reichenbach's conception, a difficulty which, as far as I know, has not previously been noted. This difficulty stems from the fact that the different values that can be substituted for the variables – for example, x – which occur in a hypothesis generally form a set with the power of the continuum, for most of those variables represent continuous magnitudes, such as the coordinates of time and of space, mass, temperature, pressure, etc. The "cases" to which a universal hypothesis refers form, consequently, a set which is not denumerable but which, in general, has the power of the continuum. On the other hand, as we have seen, in order to establish a probability statement, one has to have a well determined denumerable series of any kind of events; and the probability statement relates this series to two characteristics. Now, it is this series which, in the case of a universal hypothesis, remains completely undetermined; for, to restate the idea in an intuitive fashion, [we can say that] in contrast to a probability statement,

---

[16] Reichenbach (2), §III.

a universal hypothesis does not concern a denumerable series – but a continuum of possible cases. Thus, in the hypothesis which illustrated the form (H), "x" would be a variable whose different values represent the four-dimensional continuum of all points in space-time. If need be, one could thus consider the hypothesis in question as a set of particular propositions of the form "$F(x_i) \supset G(x_i)$";[17] however, this set would not be a denumerable series, but it would have the power of the continuum. Consequently, *universal hypotheses do not have the form of probability statements, and they do not represent sequences of individual statements.*

3. One could attempt to modify the probabilistic representation of a hypothesis by saying that the hypothesis (H) asserts, basically, that the corresponding sentence of form (HP) holds (with a certain well determined value of p) for *each* sequence that one could extract from the domain of the values of the variables which occur in the hypothesis; the hypothesis of our example would thus affirm this: for each denumerable series of spatio-temporal points $(x_1, x_2, x_3, \ldots)$, the number of points where there is iron at 1200°C and light radiation, divided by the number of those where there is iron at 1200°C, converges toward p.

Let us note first that even with the help of this interpretation, one would not be able to attribute to a hypothesis the form (HP) of a probability statement – but only a form which results from it by the introduction of a universal quantifier that refers to a variable ranging over sequences. But anyway this interpretation, which, at first glance, seems mandatory, is self-contradictory in all those cases where p is not equal to 0 or 1. In fact, let us assume that in our example p is equal to 0.9. Then it will be possible to select, in any progression of space-time points, a partial progression for which the limit of the fraction in question would be equal to 0; for that, it would suffice to select the 10% of points x for which "$F(x)$" and not "$G(x)$" hold. The assertion that p is equal to 0.9 for each progression that one can form in the domain under consideration would then be contradictory.

[17] That is not the form which is envisaged by Reichenbach and which I have discussed in Hempel (1).

4. One could try still another way of maintaining the principle by which Reichenbach wants to determine the probability of a hypothesis. One could, namely, say this: Even if a hypothesis does not represent a denumerable series of particular assertions, it remains true that we would never be able to examine more than a finite number of the cases to which the hypothesis applies. Let us then understand by the probability of a hypothesis

(H) $$(x)\{F(x) \supset G(x)\}$$

the number of those cases observed until now, where F and G were realized together, divided by the total number of the cases where F was realized. In this manner of determining, then, the probabilty of (H), one would make none of the presuppositions to which we have just raised objections.

However, this way of determining the probability of (H) has a curious property: It has the consequence that the probability of a hypothesis can change when one writes it in a different way but without changing its content.

Consider, for example, the two following formulas; they are logically equivalent representations of one and the same hypothesis, which asserts that the characteristics F and G are never realized together:

(H_1) $\qquad$ $(x)\{F(x) \supset {\sim} G(x)\}$

(H_2) $\qquad$ $(x)\{G(x) \supset {\sim} F(x)\}$

Let us now assume that the statements made so far on the subjects of the characteristics F and G confine themselves to five cases (e.g., to the observation of five objects – $x_1, x_2, \ldots, x_5$ – which can possess the characteristics in question), and that the observations are the following:

$$F(x_1) \quad F(x_2) \quad F(x_3) \quad {\sim}F(x_4) \quad {\sim}F(x_5)$$
$$\sim G(x_1) \quad {\sim}G(x_2) \quad G(x_3) \quad {\sim}G(x_4) \quad G(x_5)$$

Then the premise of (H_1) is satisfied in cases 1, 2, and 3, among which the conclusion is satisfied in cases 1 and 2; the probability would, therefore, be 2/3. On the other hand, one finds that the premise of (H_2) is satisfied in cases 3 and 5, and that the premise and the conclusion are both satisfied in case 5 – and that, therefore, the probability of (H_2) is equal to $\frac{1}{2}$.

According to the definition in question, the probability of one and the same hypothesis can thus change with the manner in which one formulates it; that makes the definition unusable. As is easily seen, these considerations highlight a difficulty faced equally by the representation, in the form (HP), of empirical hypotheses and theories.

In sum, it thus seems to me that Reichenbach's theory, as far as it concerns the probability of theories and hypotheses, cannot be maintained in its present form, even after the modifications discussed in (3) and (4); and, generally speaking, it seems that probability theory does not lend itself to the establishment of a degree of metric confirmation. Let us return, now, to the problem of empirical verification.

NO PROB. CONF.

§6. So far, we have examined only those among the empirical propositions which are capable only of an indirect verification. But one easily sees that *our considerations apply equally to directly verifiable propositions*, for each directly verifiable proposition always equally admits indirect verifications – and particularly in science one profits quite often from this possibility.

Thus, for example, instead of ascertaining by direct observation, at three o'clock in the morning, that the needle of the thermograph is between the marks 0 and −1 on the graph paper, the staff of a meteorological observatory will, in general, make use of an indirect method, which consists of examining, some hours later, the curve drawn by the needle. One applies without difficulty analogous considerations to every other case by bearing in mind that the simplest empirical affirmation has a content which goes far beyond what one can ascertain with the help of one or several direct observations – in short, that it is already a hypothesis which permits the deduction of an infinite number of consequences and to which the results apply that we have established for all propositions which are capable of an indirect verification.

This result is of particular importance for the logical analysis of verification, for the verification of any empirical proposition P rests on the deduction from P of certain consequences C which permit verification by immediate observation. But as we now see, the observation does not permit us to decide in a definitive manner whether such a consequence C is true or false. The scientific procedure of verification does not, therefore, reduce the empirical examination of a hypothesis to the examination of other propositions for which one could decide neatly as to their truth or falsity; but it does reduce the verification of a more general hypothesis such, for example, as that of Brownian motion, to the verification of less general propositions which admit of a simpler verification but which are nevertheless still hypotheses. In the exact sciences, for example, these less general propositions quite often have the following form: The needle of such and such measuring instrument will lie between such and such marks on the scale. In general, one checks a consequence C of this kind by direct observation (i.e., one establishes, with the help of appropriate observations, observation propositions O which either contradict or confirm C). The propositions of this form which one has established with the help of suitable observations thus serve in these cases as reference or basis systems in empirical verification. It is on the condition that these consequences are in good agreement with the observation propositions that a hypothesis is accepted into the system of empirical propositions that are considered as sufficiently confirmed.

But this system of observation sentences does not form an absolutely fixed and unshakeable basis, for each directly verifiable sentence is itself only a hypothesis. It can be submitted to further verifications (an observational proposition on the position of the needle of an electrometer can, e.g., be verified indirectly by a photograph of the electrometer taken at the moment of the measurement), and it is, therefore, not accepted once and for all; it is always susceptible of being eliminated from science if, one day, it does not square sufficiently well with the reference system, which is constantly modified and enlarged. (It is, by the way, the sentences adopted by science which form what Neurath calls the encyclopedia; and what we just said on the subject of methods of scientific verification can, at the same time, serve as

an answer to certain of the questions formulated in this respect by Mr. Petzäll.[18])

If one is faced by a hypothesis which has been confirmed to a certain degree by observation, then it is a question of convention to decide whether one wants to consider this degree as either sufficiently high to accept (always provisionally) the hypothesis or as sufficiently small to reject it – or whether one wants to leave the question in suspense until one has available further observations which make it possible to check the hypothesis in question. The decision concerning what is sometimes called the empirical validity of a hypothesis (i.e., the question of its adoption or rejection in science) thus contains a conventional element.[19] However, this ascertainment implies in no way the assertion that it is a matter of a completely arbitrary decision; in fact, as we have emphasized, no hypothesis is adopted without having been confirmed to a certain degree by the process of empirical verification which we have been studying, and a hypothesis, once adopted, is maintained only on condition that it be sufficiently well confirmed by subsequent verifications.

Our considerations have thus led us to abandon the absolutist conception of empirical verification and to introduce a concept of empirical confirmation which, because of its two indices, turns out to be doubly relative.

In this context, I think it important to add that the Polish logician A. Tarski has recently succeeded in establishing the theory of a concept of truth which is not relative in the sense which we have just indicated.[20] However, this theory of truth does not concern the problem of verification of which we are speaking in this section; it does not aim at determining the conditions under which one acknowledges or rejects an empirical proposition. As a truth condition for the sentence "There is a lamp on the table," for example, this theory does not give criteria of empiri-

---

[18] Petzäll (2).

[19] A. Petzäll, in his book (1), p. 23, demands sharper formulations on the subject of the subject of the conventional element, which, according to Carnap, is present in every scientific verification. The remarks made above can perhaps serve as commentary. (Hempel adds, "Refer to Neurath's work." – Editor.)

[20] See Tarski (1).

cal confirmation; it establishes rather the following criterion: The proposition "There is a lamp on the table" is true if and only if there is a lamp on the table. A theory of truth which offers such indications appears to be quite banal, but it is not so by any means. In fact, it had been believed until then that any theory of truth according to which the statement "The assertion 'p' is true" is equivalent to the sentence "p" must have as necessary consequences contradictions such as the classical paradox of the liar.[21] It is Tarski's merit to have developed a rigorous theory of truth which refutes this supposition and which, moreover, is rich in important consequences. Tarski's theory presupposes that there be given a language (which, moreover, must be formalized) – and a particular system of propositions $P, Q, R, \ldots$ expressed in this language. The theory then makes it possible to establish new propositions such as "P is true," "Q is true," etc. This theory thus does not concern itself with the question of the criteria by which one has determined or chosen the system of propositions in question; it presupposes that as given.

In particular, Tarski's theory does not concern the criteria by which the system of the propositions of the empirical sciences is established (i.e., the criteria of "empirical confirmation" which we have been studying in this section). In order to avoid any confusion about these two different problems, we have spoken, in the course of our reflections, not of "empirical truth" but of "empirical confirmation."[22] In an article which, in a very clear and instructive manner, highlights the import and the consequrences of certain of the results obtained by Tarski, Mme. Kokoszynska (1) calls the concept defined by Tarski the absolute concept of truth ("*absoluter* Wahrheitsbegriff").[23] I fear that this designation might give rise to a misunderstanding, which would consist of believing that Tarski's theory, undermining a premature criticism by the empiricists, makes it finally possible to

[21] Let us designate by "P" the following proposition which affirms its own falsity: "This proposition is false." Then one easily sees that if the proposition P is true, then it must be false; and if it is false, it must be true.

[22] This distinction has been advocated in a general fashion by Popper [(1), pp. 204–5] – and particularly in consideration of Tarski's theory by Carnap (6).

[23] *See also Tarski (1), 318.*

DON'T MISUNDERSTAND TARSKI – [CITE!]

attain the goal of so many metaphysical efforts: absolute and indubitable truths. It is hardly necessary to say that this is not the case and that Mme. Kokoszynska herself is far from wanting to assert such a thing. In fact, she emphasizes several times that the truth of a proposition in Tarski's sense is always relative to a certain language, one to which the proposition belongs, and that one and the same proposition can be true in one language and false in another (Kokoszynska, p. 155).[24] Thus, there is no question at all here of absolute truth in an ontological or metaphysical sense.

In her article Mme. Kokoszynska also raises certain objections to my article (1),[25] in which I criticized the idea of absolute empirical truth by stressing that no empirical proposition whatsoever can ever be established as absolutely true, considering that any empirical verification can attain only a more or less large confirmation. Now, Mme. Kokoszynska points out that Tarski's theory has proved that it is entirely possible to establish a concept of absolute truth. But this, I think, does not refute the assertion in question, because (1) it refers to the problem of empirical verification (in the article in question, I still used the term "truth" instead of "confirmation"[26]) and (2) Tarski's theory does not establish an "absolute" truth, as we have seen.

HEMPEL VS. KOKO-SZYNSKA [who she?]

On the other hand, I wish to emphasize that Tarski's theory reveals the possibility of establishing a concept of truth which is not based on the principle of empirical confirmation. It seems to me that each of the two concepts (*truth* and *confirmation*) clarifies and makes precise a certain facet of the ordinary and rather vague notion of truth.

Finally, Mme. Kokoszynska stresses that the concept of truth can never be interpreted as a purely syntactic concept; I am perfectly in agreement with her, and I readily admit that, in my article in question, I expressed myself in an imprecise manner on this subject. (It remains correct, however, that the concept of

[24] *Likewise Tarski himself: (1), p. 265.*

[25] See also the articles Schlick (1), (2), and (3), and Hempel (2), which touch, in part, on the same questions.

[26] At that time Tarski's investigations were not yet published in German.

"degree of confirmation of a proposition P with respect to a system S of other propositions" – or, rather, an analogous concept of a nonnumerical character – can be conceived as a purely syntactic concept.)[27]

§7. As we have already indicated at the start of these reflections, the result of our analysis – and above all the introduction of the doubly relative concept of empirical confirmation – bears also on the idea of empirical reality. In fact, if one wishes to use terms such as "empirical reality," "empirical fact," etc. – which is, sometimes, very convenient – one must in principle assign them indices of degree and of time (or of reference system), which means that the idea of empirical reality is made relative in the same sense as that of empirical verification.

But can one not at least maintain the idea of true knowledge of the absolute reality as limit-idea, as ideal? Can one not say that empirical research progressively approaches the complete knowledge of reality, without however being able ever to reach it? Such images are quite dangerous. The one we have just cited is taken from mathematics. But one knows, in mathematics, plenty of infinite sequences that have no limit; and above all, the assertion that a given sequence has a limit has meaning in mathematics only if one knows the law of formation of the sequence. In epistemology, there is nothing similar, and the idea of limit is thus undefined. It is, therefore, not a human imperfection that prevents us from ever reaching the ideal of true knowledge of reality – it is rather the fact that this ideal has no determinate content.

III. THE CHARACTER OF
LOGICO-MATHEMATICAL TRUTH

§8. Let us now proceed to an analysis of what the truth of the propositions of logic and of mathematics would be. As we have already indicated, the situation here is quite different from the one we have encountered in the domain of empirical science. In

---

[27] Compare Popper (1), p. 204.

fact, from earliest antiquity, logic and mathematics have been famous for the rigor of their reasoning, and there has been general agreement in attributing to their theorems an absolute and definitive truth. We must now consider whether this manner of considering the "formal sciences" is justified.

Let us begin by asking ourselves whether logic and mathematics are not equally of an empirical character, as certain empiricists have asserted (e.g., John Stuart Mill). According to him, the theorems of mathematics and of logic are empirical generalizations which are so well confirmed by our experiences that we are accustomed not to doubt their truth. To examine this thesis, let us consider an example. Suppose somebody comes and tells us that in such and such a hospital, a woman has given birth to two children, Charles and Henry, who exhibit the peculiarity that at the same time Charles is larger than Henry and Henry is larger than Charles. Then nobody – not even the most conscientious reporter – will put himself in the position of verifying this assertion. One would rather say without hesitation both that such a thing is logically impossible and that whoever tells us this story is either making fun of us or attributing to the notion "larger than" another sense than one usually does. Let us assume that our informant protests energetically and tells us, "I have seen with my own eyes that the doctor measured the babies twice; in the first comparison, Charles was found to be 47cm and Henry 46.5cm in length; the second measurement, carried out immediately afterward, gave 46.25cm for Charles and 46.75cm for Henry." Even then we would certainly not grant that our informant was right, but we would attribute the curious result to observational mistakes, which are always possible when one carries out measurements. In fact, observational results such as those we have considered occur quite frequently in the experimental sciences; one explains them by measurement errors or by other hypotheses, but – whatever may be the ascertainments established by observation – never would one draw from them the conclusion that, sometimes, a magnitude can be at the same time larger and smaller, or, more precisely, larger and not larger, than another. Now, what is involved here, as one will already

[margin note: ARE LOGIC + MATH EMPIRICAL?]

57

have noticed, is a case of the application of <u>the principle of excluded contradiction</u>,[28] according to which a proposition and its negation are never true together; or, more precisely: If "p" is any proposition, then the proposition "p and not-p" is always false. As our example illustrates, one maintains this principle whatever the observed empirical phenomena may be. In fact, one could not even *imagine* experiences – as extraordinary as they might be – which could lead us to consider invalidating the principle of excluded contradiction. That shows that this principle has nothing to do with our experiences – that it is, therefore, not an empirical law, as John Stuart Mill claimed, since, for every empirical law, one can imagine circumstances with which the law would be incompatible. Take, for example, the law of falling bodies. The observation – which is quite *imaginable* – that an iron ball falls faster than a wooden one would contradict that law. In other words, the law rejects the possibility of such an event – or, more briefly, the law forbids this event (and, evidently, many others as well). It is the same for every other empirical law; and it is just the fact of interdicting the realization of certain imaginable cases which gives an empirical character to a law.[29] The logical principles, however, such as that of excluded contradiction, forbid nothing. As we have already said: *whatever* the empirical phenomena may be – that physical bodies fall down or that they rise, with the same or different speeds; or that the results of certain measurements of length are such and such or, rather, others – the principle of excluded middle is not incompatible with any of these possibilities, it could not be invalidated by any imaginable event, it asserts nothing at all concerning "empirical reality," and its empirical content is zero.

One can apply without difficulty the same consideration to other logical principles such as that of excluded middle. According to this principle, any proposition of the form "p or not-p" is true, whatever the content of the proposition "p." Suppose that

---

[28] With H. Scholz (see, e.g., [1]), we prefer this cumbersome but more precise term for what is generally called "the principle of contradiction."

[29] This has been clearly put in relief by Popper in his book (1); see his § § 6 and 15.

"*p*" is the proposition "Mr. X will inherit a large fortune." Then by application of the principle of excluded middle, one arrives at the proposition "Mr. X will inherit a large fortune or he will not inherit one." This is an indubitably true proposition; we know it in advance, whatever course the life of Mr. X may take. Even theoretically it is impossible to indicate a case – however improbable it may be – in which the prediction would be disconfirmed. Thus, again, it is a matter of a proposition which asserts nothing on the subject of the future of Mr. X and which has no empirical content.

We distinguish, therefore, between logical principles, which are often called "laws of logic" (such as the principles of excluded middle and of excluded contradiction), *and the logically true propositions which one can establish with the help of these principles* (e.g., "Mr. X will or will not inherit a large fortune"). As our examples show, neither the logical principles nor the logically true propositions affirm anything about empirical phenomena; the truth [of the propositions] is completely independent of the state of our empirical knowledge, and this truth has neither an index of degree nor an index of time.

Let us proceed now to an example taken from mathematics. We will have to limit ourselves to an extremely simple case without discussing questions of detail. But we have not done otherwise for logic, for the principles of excluded middle and of excluded contradiction, to which one frequently adds a principle of identity, do not suffice, as is sometimes thought, to form the basis of the whole of logic. But these simple examples, which have the advantage of being well known, will suffice to bring out the essential points.

As an example of a true mathematical proposition, we choose this equation: $7 + 5 = 12$.

Is this an empirical proposition? Couldn't we indicate empirical circumstances which would permit it to be verified? One could, in fact, think of cases like this: We put amoebas, one beside another, first 7, and then another 5. Then, we count the amoebas in order to know whether that makes 12 amoebas in all – and we find . . . 14. Would we then say that the arithmetical proposition is false? There is no doubt – never! One would rather say that certain of the amoebas have divided in two; or else one would

59

assume some error. But never would one say that an observation has proved that sometimes $7 + 5$ can be equal to 14 – and that it has, therefore, disproved the equation "$7 + 5 = 12$." On the other hand, it is not possible either to speak of empirical confirmation: since we know in advance that the proposition "$7 + 5 = 12$" cannot be disconfirmed by any experience, it has no need of and it is even incapable of empirical confirmation, for all the methods of verification that one could imagine would never affect this equation. It has thus no empirical content. The interpretation of the principles of logic and mathematics as empirical hypotheses is, therefore, untenable.

The basic error in this interpretation is perhaps due to an inadequate[30] formulation of the logical principles in question. Thus, for example, the principle of excluded contradiction is rather often expressed in the following manner: "Something cannot at the same time exist and not exist" or "No object can at the same time possess and not possess a certain property." These formulations give the impression of being about empirical objects or facts; and in the presence of these versions, one is tempted to say: It is astonishing to see how far nature submits itself to the postulates of our logic; it never injures us by creating an object which possesses a certain characteristic – and, at the same time, does not possess it. As we have come to see, this manner of posing the problem is entirely inadequate; it makes no sense to say that empirical reality submits or does not submit to our logic and to our mathematics; for the principles of these disciplines demand nothing of "empirical reality"; they do not concern it.

§9. But then what do they concern? In a general way, the answer is this: The principles (such as that of the excluded middle), which permit us to establish propositions said to be logically or mathematically true, refer not to "empirical reality" but to the system of linguistic means by which we describe it; these principles represent rules of application which one undertakes to observe in using the language. Let us now illustrate, by a very simple example, the fundamental idea of this answer.

---

[30] It is a matter, here, of what Carnap calls the material mode of speech (*materiale Redeweise*); see, for example, (1) § §74, 78, 79, 80; (3) part III.

Suppose that a certain species of animal is described by enumerating its characteristics and that one stipulates: Each animal of this species will be called a horse, or also an equus caballus. Then one can be certain that each horse will also be an equus caballus – and inversely. It is an indubitably true assertion; but it is not a matter, here, of an empirical law, for our affirmation derives from a pure <u>linguistic convention</u>, one which stipulates that the two expressions *horse* and *equus caballus* will be synonyms. It tells us nothing about the subject of the species *horse* or *equus caballus*. It would, therefore, make no sense to say: How astonishing it is to see how far nature bows to our definitions; it never produces a horse which is not at the same time an equus caballus. In the same way, the expressions "7 + 5" and "12" of our arithmetical example are synonymous because of a convention which consists in adopting our system of numerals as well as the principles of our arithmetic; this convention permits us to develop, for a given numeral, even an infinity of symbols which are its synonyms.

The situation is basically the same for all propositions which we call logically true; their truth derives in the same way from conformity to certain conventions concerning the use of the language of science.

For some time, now, we have been able to determine in an exact manner the character and content of these conventions – and thus to establish a precise definition of logico-mathematical truth. We owe this result to recent research in the domain of what is called logical syntax; these investigations depend on methods and results of logistic, the new symbolic logic which treats the problems of logic with the help of a calculus that resembles the calculi of mathematics and which confers upon logical investigation a rigor far surpassing that of traditional logic.

Without being able to enter into details[31] of these very

---

[31] The theory of syntax has been developed in detail by R. Carnap in his work (2a, 2b); in a more elementary form, the fundamental ideas of this theory are presented in Carnap (3). For the analysis of mathematical truth, compare also Carnap (5). Carnap (2a, 2b) contains very complete references to the writings of other authors who have contributed to the solution of the problem in question; these are above all K. Gödel and A. Tarski.

complicated investigations we offer the following exposition, which tries to characterize in a general and intuitive way at least the fundamental ideas which emerge from the established results.

The investigations in question concern the most general characteristics of the formal structure of a language; they develop, as one says, the general syntax of any language whatsoever. One designates, in these investigations, as a "language" not only every living or dead language, but also, first, the most precise system of linguistic representation which physics has developed; furthermore, all formal systems, such as the mathematical theory of vectors; and, finally (and even most of all, because of their formal clarity), the formalized "languages" which have been constructed expressly by logistic (e.g., the celebrated system of Russell and Whitehead's *Principia Mathematica*).

The logical structure of a language is completely determined by three groups of characteristics, which we will call, in a general way, the *formal determinations of the language*. These are: its vocabulary, its rules of formation, and its rules of transformation.

The *vocabulary* enumerates, without indicating their meaning, all the elementary symbols of which the complex expressions of the language are composed (e.g., the propositions). The elementary symbols of ordinary language are the words, the number symbols, the punctuation signs, etc. In the "language" represented by algebra, these are the symbols which enter into the algebraic formulas.

The *rules of formation* stipulate which series of elementary symbols are counted as propositions, without, however, making any distinction between true and false propositions. Thus, for example, for the language of arithmetic, the series of symbols "$7 + 5 = 12$" as well as the series "$7 + 5 = 14$" will be counted as propositions by the rules of formation (that is to say, as expressions which have the form of a proposition); whereas the expression "$+^2 = 4$" will not be counted as a formally legitimate proposition, even though each of the symbols composing it is an elementary symbol of arithmetic.

The *rules of transformation*, finally, determine the principles of logical deduction;[32] these are the principles by which one can deduce from a class of propositions that form the premisses a proposition called a logical consequence of these premisses. A system of rules of transformation of a language can be formulated as a system of stipulations whose totality defines the notion "logical consequence."[33] That is to say, one can give each rule of transformation the following form: "Given a class of propositions having such and such forms, then a proposition of such and such another form is a consequence." Thus, for example, the rule corresponding to classical modus ponens can be expressed in the following form: Given a class of two propositions having the forms "p or q" and "not-p" (where "p" and "q" are any propositions of ordinary language), then "q" is a consequence. As one can see, this rule can be expressed in a purely formal manner; it does not presuppose the truth of the premisses; it does not refer to the content of the propositions at all, but only to their form. In applying this rule to a particular case, one may find, for example, that the class of the two propositions about a certain basset hound ("Fifi is brown or is black" and "Fifi is not brown") has as a consequence the proposition "Fifi is black."

Let us note that the *general* principles which are usually called logical laws (e.g., the principle of excluded middle, of excluded contradiction, etc.) equally represent transformation rules, which by the way have a curious form. The principle of excluded middle, for example, in a correct form, says this: Every class of premisses, even the empty class, possesses as a consequence each proposition of the form "p or not-p," where "p" is any proposition whatever. This expresses, to use more intuitive terms, that the validity of propositions of the indicated form does not depend on any premise but that they are unconditionally true.

The principle of excluded middle is thus a transformation rule, and it is the same with the other principles which are

---

[32] *But see footnote 4.*

[33] One can also establish empirical rules of transformation (which Carnap calls *P-Bestemmungen*); for greater simplicity, we consider here only the case where each rule of transformation is of a purely logical character (and thus constitutes an *L-Bestimmung* according to Carnap). For details of this distinction, compare Carnap (2a and 2b), §51; (3), II, 5.

ordinarily called "laws of logic." This shows that it is the system of transformation rules – and, thus, solely the definition of the notion of consequence – which determines the entire logic of a language.

Before we continue, let us illustrate by a comparison the system of formal determinants of a language. We can compare a language and the logical operations to which one submits its propositions to the game of chess. To the different elementary symbols correspond the different pieces; to the formally legitimate sentences correspond the conceivable distributions of pieces on the chess board; to the series of symbols which do not form propositions correspond the distributions which "have no sense" in the game (e.g., a distribution in which a piece would occupy several squares). The transition from a class of propositions to one of its consequences will be represented by a move which transforms one distribution of pieces into another. It is, therefore, the rules concerning the movements of the pieces – the rules of chess, properly speaking – which correspond to the rules of transformation relative to a language.

Now, if one knows only the stipulations of which we have just spoken, then one evidently will not be able to predict the development and the outcome of a game between two chess players. But, nonetheless, one will be able to make certain assertions which are valid for every game of chess, whatever its individual particularities and whatever its outcome. Thus, for example, one can assert that in any game of chess which is played in accordance with the rules of the game, it cannot happen that the two white bishops are on squares of the same color. This is an assertion which holds for every game of chess, whatever course it may take, for, as one easily sees, this assertion follows simply from the rules which one undertakes to observe.

The situation is completely analogous in the case of the "game of language." If one knows only the list of elementary symbols as well as the rules of formation and of transformation, one can evidently not predict the events which will occur – that is to say, one does not know from that knowledge the propositions which will be empirically confirmed. But, again, using nothing but the formal rules, one will be able to make certain statements which will be valid no matter what propositions one accepts as confirmed by experience. One will be able to say, for example, that in no case will one be led to establish propositions such as "Charles is at once larger and not larger than Henry." No empirical verification is needed to know that such a proposition is false. Exactly as in the case of the two bishops on squares of the same color, it is, here again, the rules of the game which exclude this proposition, which we shall call logically false. Generally speaking, the propositions which are called logically or mathematically false are those which are already excluded by the formal rules of the language; and the propositions called logically or mathematically true are those which one can establish, not by making any experiment or observation, but by means of only the formal rules of the language. This is the case, for example, for the proposition "Mr. X will inherit a large fortune or he will not." The truth of this affirmation is purely formal, and it derives from the rules of the game of our language, which have, as a consequence, the assertion of any proposition of the form "p or not p." In contrast to the empirical propositions, the logically or mathematically true propositions affirm nothing about any object; their validity is only a consequence of the formal rules which one agrees to observe in making use of the language in question. Mathematical or logical truth and, equally, falsehood have thus a purely formal character; we can also speak of the formal truth or falsity of a proposition.

The last indications serve simply to illustrate, in a manner more intuitive than precise, the basic idea of the exact definitions at which the syntactical investigations previously mentioned have arrived. Let us now glance at these definitions.

That of formal truth is the following: A proposition is called formally true or *analytic*, if it is a logical[34] consequence of any class of propositions. As we have noted in interpreting the principle of excluded middle as a transformation rule, this definition basically expresses only the idea that the truth of an analytic proposition is independent of any premise, that an analytic proposition is true without conditions. Thus, for example, the principle of

[34] Hempel adds this word and refers to added footnotes 4 and 32. (Editor)

excluded middle stipulates that any proposition of the form "*p* or not-*p*" is a consequence of any class of propositions; hence, by the definitions which we have introduced, any proposition of this form is an analytic proposition.

The precise definition of "formally false" is the following: A proposition is formally false or contradictory if every proposition is a logical consequence of it. This corresponds to the circumstance that, once a logical contradiction has slipped into an argument, one can deduce from it any consequence whatever. In more intuitive terms, we might say: A contradictory proposition is false without conditions; this is the case, for example, for any proposition of the form "*p* and not-*p*."

Finally, propositions which are neither analytic nor contradictory are called _synthetic_; these are also the so-called empirical propositions (the confirmed as well as the nonconfirmed) of the language under consideration.

§10. It is, perhaps, useful to emphasize here that *the modern construction of the concepts "analytic" and "synthetic"* which we have sketched is _not_ *identical to the definitions established by Kant.* The Kantian definition stems from the classical conception of a predicative logic; according to this any judgment consists in the attribution by a "predicate" of certain characteristics to a "subject"; and Kant said that a judgment is analytic if the idea expressed by the subject already includes the characteristics which are attributed to it by the predicate. Such is the case in the judgment "every horse is an animal." Now, logistic research has shown that the structure of most languages is not predicative in the sense presupposed in the Kantian definition. Thus, for example, the equation "$2 \cdot 3 = 6$," which is a proposition of ordinary language as well as of the arithmetical language, has no predicative structure, any more than does the physical law of falling bodies or the proposition "If *a* is larger than *b*, *b* is not larger than *a*." The first and last of these three propositions are, incidentally, analytic, without, however, satisfying the Kantian criterion, since they do not possess a subject or a predicate in a precise sense; they have not a predicative structure but a relational structure. The classical definition of the concept *analytic* hence can only be applied to languages having a quite peculiar

66

propositional structure which is neither that of ordinary languages nor that of the languages of physics and mathematics. On the contrary, the new definition of the concept *analytic*, which is provided by logical syntax, makes no presupposition concerning the formal structure of propositions; it is applicable to any language for which the formal constructions – and, in particulary that of "consequence" – are fixed.

A second essential difference between the Kantian and syntactical definitions of the concept *analytic* is this: The classical definition refers to the *meaning* of the terms of which the given proposition is formed. It speaks of the inclusion of the idea of the predicate in that of the subject – or of similar relations. The syntactical definition on the contrary is completely formal, for it refers only to the notion of consequence, which, itself, is defined in a purely formal manner by the rules of transformation. This formalization brings with it several advantages for the elucidation of the character of logic and mathematics.

One of these advantages consists of eliminating all psychologism concerning logic. The nonformal definition of the logical concepts, such as the Kantian definition of the concept *analytic*, has brought with it questions such as: In representing to oneself the idea of the subject of such and such a proposition, does one represent to oneself at the same time the idea of its predicate? In fact, Kant himself characterized the proposition "$7 + 5 = 12$" as analytic, on the basis of the following reasoning, which is, it seems, influenced by this manner of seeing the problem:

One is surely tempted to believe at first that this proposition $7 + 5 = 12$ is a purely analytic proposition, which follows by the principle of contradiction from the concept of the sum of seven and five. But if one looks closer, one finds that the concept of the sum of 7 and 5 contains nothing more than the union of the two numbers in a single one, and that we have no knowledge of which particular number that one may be. The idea of twelve is not at all conceived simply in my conception of this union of 5 and 7, and no matter how well I might analyze my concept of such a possible sum, I would not find the number twelve in it . . . . In the concept of sum $= 7 + 5$, I have clearly recognized that 7 should be added to 5, but not that this sum was equal to 12. Arithmetical propositions are thus always synthetic; that is what one will see still more clearly when one takes larger numbers: it then becomes evident that no matter how carefully we look at our concepts, we will never find the sum by mere analysis, without resorting to intuition. (Kant, *Critique of Pure Reason*, 2nd Ed.) [1787] Introduction, pp. 16–17.

It is these elements of a psychological order which slip into Kant's thought, and the question whether a given proposition is analytic or not leads him, thus, to a conclusion which, in the light of a rigorous examination, cannot be maintained. In the formal version of the concept *analytic* (and of other concepts of logic), this danger is avoided. If the formation and transformation rules of a language are expressed in a purely formal manner, it is the same for the definition of the conceps *analytic, contradictory*, and *synthetic*; and one can, in principle, using only these definitions, distinguish a synthetic proposition from an analytic proposition (or from a contradictory proposition) without knowing the meaning of these propositions – even without knowing the meaning of any of the elementary symbols of the language. Likewise, one can establish logical deductions without knowing the meaning of the propositions which occur in them. This, incidentally, is a result of the greatest importance for the modern investigation of the foundations of logic and mathematics.

§11. Let us now return to the concept of an analytic proposition. We have said that the validity of analytic propositions is not based on the formal rules of the game, which consist of using a certain language; and we have compared these rules to those which govern the game of chess. Now, it seems that, all the same, there is a fundamental difference between these two systems of rules. The determination of the rules of a game is notably arbitrary; we can choose them according to our wishes, and we are always free to modify them at our pleasure. For the rules of the game of language, the case seems absolutely different. In fact, it seems that we cannot choose as we please between two imaginable systems of logical deduction, of which

68

one contains the principle of excluded middle, whereas the other does not. In fact, it is incontestable that, in ordinary language, we generally conform to the principle of excluded middle and other syllogistic principles of classical logic to such a degree that it seems to us impossible to follow any other system of rules; but from that, one cannot yet conclude that there could not exist languages which do not contain these principles. On the contrary, modern logic has now established, even in great variety, languages of that description which are called non-Aristotelian – that is to say, systems whose transformation rules do not comprise certain principles of classical logic, especially the principle of excluded middle; these systems, nonetheless, imply no absurdity. We owe these system, above all, to the investigations of J. Lukasiewicz and his school, of E. L. Post, of C. I. Lewis, as well as of the school of intuitionist mathematicians.[35] The development of these non-Aristotelian logics has only been possible thanks to the formal methods of which we have spoken. One represents the propositions of the system that is to be constructed by logistic symbols; one formulates the desired rules of transformation; and one "makes one's move" without thinking of the possible meaning of the results. It is in this way that one avoids the danger of sliding into the deductions of intuitive reasoning – of modes of inference forbidden by the rules of the game – and it is this method which has made possible the rigorous demonstration of the possibility of different non-Aristotelian logics.

The idea of a logic other than that which governs the ordinary languages thus turns out to be neither absurd nor inconceivable; but neither can we say either that a non-Aristotelian logic is false, for, in setting up a language, one can give it the formal structure one desires. Its formal specifications can no more be true or false than the rules of a game, but they can be more or less convenient in a certain context (e.g., more or less well adapted to the needs of certain empirical science).

That holds equally for those among the formal stipulations

---

[35] Compare Lukasiewicz (1), Lukasiewicz and Tarski (1); Post (1); Lewis (1); Lewis and Langford (1); Becker (1); Heyting (1).

which determine the mathematics of the language in question. In particular, arithmetic need not necessarily be based on our infinite system of whole numbers with their characteristic properties; and one could even imagine an empirical world such that, for its description, a language containing "our" arithmetic while not being "false" would perhaps be less convenient than a language including a quite different arithmetic (e.g.,[36] a system like that which is used by certain primitive peoples and which contains only the numbers "1," "2," "3," "4," and "many"). If, for example, our experience were such that we could not isolate quite distinctly relatively persistent objects – but that all we perceived were to glide rather quickly like the images of falling drops – then it would perhaps be very convenient to use this other arithmetic which is simpler than ours. But certainly it would not be false even though its theorems do not correspond to those of our arithmetic – nor vice versa.

§12. Let us now draw the *conclusions* of our considerations to see whether the thesis of the absolute and unshakeable truth of logic and mathematics is justified.

It turns out that mathematical formulas and theorems, along with the propositions which are said to be logically true, are analytic propositions, which are unshakeably true in the sense that no experience and no phenomenon, however unexpected and improbable, can ever disconfirm them. But this truth is quasi-sterile, for it is due to the fact that analytic propositions assert nothing about any object, not even about any ideal objects, as has often been pretended in philosophy; more precisely, the truth of analytic propositions stems from the rules which we ourselves agree to follow when employing the language to which these propositions belong.

And yet, this purely formal truth is not absolute: for it depends essentially on the formal specifications which we impose on the language in question. One cannot, therefore, say, "Such and such a proposition is analytic," but only, "It is analytic relative to a language which has such and such syntax." What are called

---

[36] The following example is taken from the recent book of Waismann (1), which gives an excellent introduction into the logical analysis of mathematics.

mathematical and logical truths hold, therefore, only in a system of reference whose specification has the character of a convention. The choice of the syntax which determines the logic and mathematics of a language can be regulated not by truth criteria but only by practical considerations such as the question of whether the form of the language is convenient in relation to the context for which the language is intended.

But once the formal system of reference is fixed, the determination of the logically and mathematically true propositions is no longer free: It is unambiguous criteria which determine the formally true propositions of the adopted system. The propositions at which we thus arrive teach us nothing. There lies their weakness – but at the same time their strength, for that confers upon them, in the system in question, a definitive and general truth which is due to the fact that these propositions only make explicit the reach of our own conventions.

IV. CONCLUSION

§13. We thus arrive at the result that neither the empirical nor the formal sciences can lead us to a truth which is absolute and independent of our thought. It might seem, at first glance, that this represents a great theoretical renunciation, that this is the confession of a limitation in principle concerning the possibilities of human knowledge. But that is absolutely not the case. For one cannot speak of a theoretical renunciation except in regard to a problem with a well determined meaning, one which is thus posed in such a way that, for a proposed solution, one can always decide whether or not it resolves the problem.

But this condition is just not satisfied by the problem of finding absolute truths. It is impossible to imagine a criterion which allows us to decide whether a putative truth is an absolute truth or only an apparent truth, for whatever the proposed criterion, one could always doubt whether it is a true criterion which exclusively determines absolute truths. Therefore, the result we reach about the question of absolute truth is not that we must give up all attempts *to solve a genuine, well determined problem*, but that we have succeeded in *dissolving a pseudoproblem*.

BIBLIOGRAPHY

Becker, O., (1) "Zur Logik der Modalitäten," *Jahrbuch für Philosophie und phänomenologische Forschung* 11 (1930), 497–548.

Carnap, R., (1) "Über Protokollsätze," *Erkenntnis* 3 (1932), 215–28.

(2a) *Der logische Syntax der Sprache* (Vienna: Julius Springer, 1934).

(2b) Amethe Smeaton, Countess von Zeppelin, trans., *Logical Syntax of Language* (London: Kegan Paul, 1937). [Enlarged translation of (2a), revised by Olaf Helmer.]

(3) *Philosophy and Logical Syntax* (London: Kegan Paul, 1935).

(4) "Formalwissenschaft und Realwissenschaft," *Erkenntnis* 5 (1935), 30–7.

(5) "Über ein Gültigkeitskriterium der Sätze der klassischen Mathematik," *Monatshefte für Mathematik und Physik* 42 (1935), 163–90.

(6) "Wahrheit und Bewährung," *Actualités Scientifigues et Industrielles* 391 (Paris: Hermann & Cie, 1936), 18–23.

(7) "Testability and Meaning," *Philosophy of Science* 3 (1936), 419–71; and 4 (1937), 1–40.

Grelling, K., (1) Review of Popper (1), in *Theoria* 3 (1937), 134–9.

Hahn, H., (1a) *Logik, Mathématik und Naturerkennen* (Vienna: Gerold, 1933).

(1b) *Logique, mathématique et connaissance de la réalité* (Paris: Hermann, 193[4?]. Translation of (1a).

Hempel, C. G., (1) "On the Logical Positivists' Theory of Truth," *Analysis* 2 (1935), 49–59. (Chapter 1 here.)

(2) "Some Remarks on 'Facts' and Propositions," *Analysis* 2 (1935), 93–6. (Chapter 2 here.)

(3) Review of Popper (1), in *Deutsche Literaturzeitung* 58 (1937), 309–14.

Hempel, C. G., and P. Oppenheim, (1) *Der Typusbegriff im Lichte der neuen Logik* (Leyden: Sijthoff 1936).

Heyting, A. (1) "Die formalen Regeln der intuitionistischen Logik," *Sitzungsberichte der Akademie der Wissenschaften in Berlin, Sitzungsberichte der deutschen Akademie der Wissenschaften zu Berlin* (1930), 57–71, 158–69.

Kokoszynska, M., (1) Über den absoluten Wahrheitsbegriff und einige andere seantische Begriffe," *Erkenntnis* 6 (1936), 143–65.

Lewis, C. I., (1) "Alternative Systems of Logic," *The Monist* 42 (1932), 481–507.

Lewis, C. I., and C. H. Langford, (1) *Symbolic Logic* (New York: Century, 1932).

Lukasiewicz, J., (1) "Philosophische Bemerkungen zu mehrwertigen Systemen des Aussagenkalküls," *Comptes Rendus des séances de la Société des Sciences et des Lettres de Varsovie* 23 (1930), 51–7.

Lukasiewicz, J., and A. Tarski, (1) "Untersuchungen über den Aussagenkalkül," *Comptes Rendus des séances de la Société des Sciences et des Lettres de Varsovie* 23 (1930), 30–50.

Neurath, O., (1) "Physicalism," *Monist* 41 (1931), 618–23.

(2) "Physikalismus," *Scientia* 50 (1931), 297–303.

(3) "Soziologie im Physikalismus," *Erkenntnis* 2 (1931), 393–431.

(4) "Protokollsätze," *Erkenntnis* 3 (1932), 204–14.

(5) "Radikaler Physikalismus und 'wirkliche Welt,'" *Erkenntnis* 4 (1934), 346–62.

(6) "Pseudorationalismus der Falsifikation," *Erkenntnis* 5 (1935), 353–65.

(7) *Le Dévelopment du Cercle de Vienne et l'Avenir de l'Empirisme Logique* (Paris: Hermann, 1935).

(8) "Physikalismus und Erkenntnisforschung," *Theoria* 2 (1937), 234–7.

(9) "L'encyclopédie comme 'modèle,'" *Revue de Synthèse* 12 (1936), 187–201.

Petzäll, A., (1) *Zum Methodenproblem der Erkenntnisforschung* (Göteborg: Elander, 1935).

(2) "Physikalismus und Erkenntnisforschung," *Theoria* 2 (1937), 232–4.

Popper, K., (1) *Logik der Forschung* (Vienna: Springer, 1935).

Post, E. L., (1) "Introduction to a General Theory of Elementary Propositions," *American Journal of Mathematics* 43 (1921), 163–85.

Reichenbach, H., (1) *Wahrscheinlichkeitslehre* (Leiden: Sijthoff, 1935).

(2) "Über Induktion und Wahrscheinlichkeit," *Erkenntnis* 5 (1935), 267–84.

Schlick, M., (1) "Über das Fundament der Erkenntnis," *Erkenntnis* 4 (1934), 79–99.

(2) "Facts and Propositions," *Analysis* 2 (1935), 65–70.

(3) *Sur le fondement de la connaissance* (Paris: Hermann, 1935). [Containing articles (1) and (2), among others, in French translation.]

Scholz, H., (1) *Geschichte der Logik* (Berlin: Junker and Dünnhaupt, 1931).

Tarski, A., (1) "Der Wahrheitsbegriff in den formalisierten Sprachen," *Studia Philosophica* I (1935), 261–405. See also Lukasiewicz and Tarski (1).

Waismann, F., (1) *Einführung in das mathematische Denken* (Vienna: Gerold, 1936).

Chapter 5

The Irrelevance of the Concept of Truth for the Critical Appraisal of Scientific Theories

1. INTRODUCTION

Science is often said to be a search for truth, a quest aiming at the formation of true beliefs about our world – beliefs concerning particular facts was well as general laws that may connect them.

But however appealing this conception may seem, it has, first of all, fundamental logical flaws, and second, it fails to do justice to a group of considerations that govern the critical appraisal and the acceptance or rejection of hypotheses and theories in science.

I propose briefly to elaborate this claim and to suggest an alternative way of characterizing science as a goal-directed endeavor.

The following considerations are informed both by the ideas of logical empiricism and by the more recent methodological explorations in a pragmatic-sociological vein undertaken by Thomas Kuhn and by kindred thinkers.

2. SCIENTIFIC THEORIZING: INVENTION AND CRITICAL APPRAISAL

Scientific theories are introduced in an effort to bring order into the diversity of the phenomena we encounter in our experience. How are such theories arrived at, and how are they supported?

Devising of effective theories is not a matter governed by systematic rules of discovery. As Einstein was fond of saying,

75

scientific hypotheses and theories are arrived at by free invention, by the exercise of a creative scientific imagination. But their actual acceptance in science is subject to a critical evaluation by reference to the results of experimental or observational tests and to certain important additional criteria, which will be considered shortly.

To appraise the view that scientific theorizing is systematically directed toward the attainment of truth, let us briefly consider the principal standards that inform the critical appraisal of scientific theories, and let us see what bearing the selection of theories in accordance with those standards has on the question of whether the accepted theories are true.

3. THE EMPIRICIST REQUIREMENT: TESTABILITY AND EVIDENTIAL SUPPORT

There is agreement in science about one basic requirement, which I will call the principle of empiricism: The hypotheses and theories advanced in empirical science must be capable of test by reference to evidence obtained by observation or experimentation. This condition applies to all areas of scientific inquiry. Even the most esoteric theories of cosmology must in the end be testably linked to potential empirical evidence – and so too must biological, psychological, sociological, and historiographic hypotheses. Otherwise, those theories have no bearing on empirical phenomena.

Schematically, testability of the kind here called for might be viewed as the possibility of logically deducing, from the hypothesis or theory under test, the outcomes of suitable observations or experiments. Actually, the logic of the situation is, as a rule, much more complex, as was emphasized already by Duhem early in this century. In order to deduce observation statements from a hypothesis, one has to combine the latter with a large set of further assumptions concerning the experimental setup, the structure and theory of the apparatus used, the state of the environment, and the like. Hence, if the actual test findings conflict with the observation statements deduced from the theory, that shows at best that some of the premisses used need revision –

but not that this has to be the hypothesis or theory under test. This consideration leads to a holistic view of scientific knowledge in the spirit of Quine, according to which tests and other modes of critical appraisal always concern, in principle at least, the totality of beliefs held at the time.

For our present purposes, however, the simple deductivist conception of empirical testing will suffice. Our question is, "What can favorable evidential data obtained by empirical tests show about the truth of a scientific hypothesis?" Surely, they cannot, in general, furnish conclusive proof; that is the crux of the problem of induction. For example, evidence to the effect that all instances of kind A so far examined have had the characteristic B does not conclusively prove the hypothesis that all A are B.

And just as there are large classes of hypotheses that cannot be conclusively proved by empirical findings, so too there are large classes of scientific hypotheses which are incapable of conclusive disproof or refutation by negative observational data. This holds, for example, for hypotheses of existential form, such as that there are pink elephants, that there are black holes in the universe, or that there are particles traveling faster than light. In short, infinitely many important kinds of scientific hypotheses are, for purely logical reasons, incapable of conclusive proof or disproof – verification or falsification – by experimental test data. To say that scientific inquiry aims at finding truths on the basis of experimental evidence is, therefore, to characterize science as an enterprise whose goal it is logically impossible to attain.

But the conception of science as a search for truth is strictly untenable for yet another reason. Let us note that there actually are certain cases in which a hypothesis can be conclusively proved or refuted by observational evidence. Suppose, for example, that an observer, upon examination of a given bird, reports, "This bird is a white raven." This observation statement clearly affords a conclusive proof for this hypothesis: "There are white ravens." But what that proof shows is only that *if* the premise – that is, the observation report – is true, *then* so is the conclusion that there are white ravens. Thus, the proof establishes the truth of the hypothesis not categorically, but only conditionally – namely on condition that the observation report serving as premise is true.

In sum, the question whether a hypothesis is true is not even *touched* by the most favorable test results; the outcome of empirical tests has no bearing on the truth of a hypothesis.

Note that in a deductively conclusive proof of the kind just considered, the premise – that is, the scientific observation report – is by no means always true. On the contrary, the statements made by scientists about their observational or experimental findings are all subject to observational error and to distortion by various disturbing factors, not to mention the possibility of deliberate misrepresentation. Besides, even the results obtained by careful observation or measurement are sometimes set aside in science if they conflict with a comprehensive and so far widely successful theory. As Whitehead said, good theories are not as easy to come by as blueberries.

To repeat, then: Satisfaction of the empiricist requirement that scientific theories be testable and that their test implications conform to empirical evidence has no bearing at all on the question of their truth. The requirement may be satisfied by a theory that is false, and it may fail to be satisfied by a theory that is true.

In the tradition of logical empiricism, the idea was developed in great technical detail that while empirical data cannot, in general, provide conclusive verification or falsification for a hypothesis, they may provide more or less strong evidential support, or confirmation, for it, and that this degree of support may be viewed as the degree of rational credibility, or of inductive probability, which the given evidence confers upon the hypothesis. Carnap, in particular, developed an ingenious, precise theory of what he called logical or inductive probability.

But that probability is always defined as a relative notion: $c(h,e)$ is the degree to which hypothesis h is confirmed by evidence e. Such probability clearly is not a semantic feature. Probability is not partial truth; high probability is not closeness to truth. A hypothesis with high probability relative to the total available evidence may be false; one with low probability may be true.

Probability, like scientific preferability or acceptability for hypotheses, is not a semantic-ontological concept, but a relativized epistemological one.

4. DESIDERATA FOR THEORY CHOICE

In sum, then, adherence to the empiricist requirement contributes in no way to a search for truth. The requirement calls, in fact, for a quite different conception of the directedness of scientific research. Such a different conception of "the goal" of science is called for also by further considerations which were often mentioned in earlier methodological writings but which have been elaborated with particular force in the pragmatically oriented literature of recent decades, of which Kuhn's work on the structures of scientific revolutions is a prominent exemplar.

The new pragmatist turn in methodology has focused attention on the fact that when it comes to the critical appraisal and to the acceptance or rejection of proposed hypotheses and theories, a variety of considerations are brought into play in addition to the empiricist concern with testability. These considerations point to certain characteristics that count as highly desirable features of scientific theories. Such marks of theoretical merit are often referred to as desiderata. They may be viewed as reflecting, broadly speaking, the direction in which scientific research is headed, the objectives at which it is aiming.

1. First among them, there is the empiricist desideratum, just discussed, of testability and evidential support. Somewhat more fully, we might construe it as calling for a clear and logically consistent formulation of a theory, one yielding implications that are in suitable accord with evidence provided by observation and measurement. This desideratum does play an important role in the appraisal of scientific theories; but, as we saw, it has no bearing on the question of truth.

Note that logical consistency is called for, because an inconsistent theory implies any conceivable observational prediction as well as its negation and thus tells us nothing about the world.

79

The demand for agreement of test results with test implications derived from a theory can be very exacting when the theory is expressed in quantitative terms and the implications are stated with high quantitative precision. Experimental confirmation can then carry much weight.

And yet, the mere fact that the test implications of a hypothesis agree with all the available evidence does not, strictly speaking, amount to very much. That is shown by the following consideration. Suppose that we are concerned with quantitative hypotheses of a simple kind which represent the length of a given metal bar as a mathematical function of its temperature; and suppose further that we are given a finite or even denumerably infinite set of test data listing pairs of associated measured values of temperature and length. Then there exist infinitely many different hypotheses, each of them presenting the length as a precise mathematical function of the temperature, such that each of them fits all the data exactly and yet any two of the hypotheses are logically incompatible with each other.

The desideratum of testability and of conformance with the available test data thus surely does not constrain the set of admissible hypotheses very much; and further desiderata – with further modes of appraisal – do indeed enter into the process of theory choice in science.

2. One of them is the desideratum of large scope: Science prizes theories that cover a wide variety of occurrences, and science permits the prediction of phenomena that were unknown – or at any rate not taken into account – when the theory was devised.

But the preference given to theories of large scope, of great boldness, surely does not reflect a quest for truth. Indeed, if one of two theories has a larger scope than the other – in the strict sense that the first logically implies the second, but not conversely – then, no matter what the evidence, the former has a smaller probability of being true than does the latter.

3. The famous desideratum of simplicity is another important factor that enters into the critical appraisal of hypotheses and theories. Other things being equal, the simpler of two competing theories is to be given preference. Though no one precise and

[VERY IMPT.]

[I.E. WHAT HE'S SAYING HERE IS THAT WHAT DRIVES SCI. PRACTICE IS NOT QUEST FOR "TRUTH": COMPARE THIS TO 17th-CEN. SCI. WHO SAYS HE STUDIES NATURE TO "HONOR GOD,"

generally accepted characterization of simplicity is available, scientists often agree when judging the relative simplicity of proposed conjectures, such as a smooth curve through a set of data points, in contrast to a curve that contains various irregular loops, or the oxygenation theory of combustion, as contrasted to the late forms of the phlogiston theory. The desideratum of simplicity thus does play a considerable role in the critical appraisal of theories, but its satisfaction clearly has no bearing on the question of their truth.

4. The eminent theoretical physicist and Nobel laureate Paul Dirac gave the preference for simplicity a provocative new turn: He held that "the basic equations of physics have to have great mathematical beauty." Such beauty, he claimed, does not depend on cultural or personal factors: "It is the same in all countries and at all periods of time." Although he offered no precise definition of mathematical beauty, he attributed to it a fundamental role in the appraisal of physical theories. In particular, he held that if an experimental finding conflicts with a mathematically beautiful theory, then the experimental finding has to be questioned.

But why should simplicity – or why should mathematical beauty – be considered to be such an important theoretical virtue?

I once had an opportunity to ask Dirac why he regarded mathematical beauty as an essential mark of a sound physical theory. His answer: Because the world itself is basically beautiful – God made it so.

There are those who cannot accept the metaphysical assumption presupposed by this answer. I count myself among those skeptics.

I am inclined to agree instead with a pragmatically oriented conception according to which the importance of simplicity as a theoretical virtue reflects, at least in part, certain limitations of the human mind in its search for comprehensive accounts of the world.

For success in that search, it is essential to devise theoretical systems that are simple enough to be conceived and applied with the limited observational and conceptual resources that human

[A BIOLOGICAL-PRAGMATIST ARGUMENT FOR THE SIMPLICITY CRITERION!]

investigators are endowed with. Advances in theoretical simplicity and conceptual integration thus enhance our ability to
understand the world and, to some extent, to anticipate its
changes.

5. EPISTEMIC OPTIMALITY

We have surveyed some of the leading considerations that enter
into the critical appraisal of scientific theories and into decisions
about their acceptance or rejection. We found those considerations to be aimed at meeting various desiderata whose fulfillment
by a theory has no bearing on the question of whether the theory
is true.

So, if scientific inquiry is to be viewed as a systematically goal-
directed enterprise, its goal surely is not the attainment of true
theories. Our considerations about theory choice in light of the
desiderata rather present scientific theorizing as directed toward
the construction of well-integrated world pictures that optimally
incorporate our experimental data available at the time into a
simply and smoothly cohering and far-reaching conceptual
scheme. Scientific theorizing is oriented, we might say, not at the
ontological goal of truth but at the epistemological one of optimal
epistemic integration – or at epistemic optimality – of the belief
system we hold at any time.

The concepts of truth and of epistemic optimality differ fundamentally. Each plays an important role in science: the former
as a basic concept of logic and semantics, the latter in the theory
of knowledge and the methodology of science. They have quite
different characteristics.

For example, the concept of truth for descriptive statements in a
given language is absolute and timeless. A statement is either
true or false, independently of any evidence, and its truth value
– that is, truth or falsehood – does not change in the course of
time. If a statement is not true, then its negation is true. Any
logical consequence of a true statement is true – and so on.

The epistemic value of a theory, on the other hand, is subject
to change. It is relative to – and dependent on – the epistemic

"TRUTH" vs.
"EPISTEMIC VALUE"/"EPIS. OPTIMALITY"

situation at the time, including the empirical data then available and whatever competing theories may be under consideration for appraisal. The temporal changeability of epistemic value is clearly reflected in the process of theory change in science.

And there are further contrasts to the characteristic of truth just mentioned. For example, it is not the case that if a theory is not epistemically optimal, then its negation is. Both may be irrelevant to the epistemic situation at hand. A logical consequence of a theory of great epistemic value may be so weak in content as to be scientifically worthless – and so on.

To say that scientific inquiry is oriented toward the construction of world pictures that are epistemically optimal in the sense just outlined is clearly an oversimplifying schematization. The desiderata governing the critical appraisal and the adoption or rejection of theories in science do not form a sharply delimited set, and even the very important ones we have considered are not amenable to precise formulation.

Similar difficulties arise for an attempt to indicate the relative weight of the desiderata in theory choice. Suppose that one of two competing theories satisfies certain desiderata particularly well, but suppose it is rather weak in relation to certain other desiderata, which are very well met by the second theory. How are those advantages and deficiencies to be compared and weighed? To which of the contending theories do they accord the higher theoretical merit?

Yet all these indefinitenesses do not leave theory choice a purely subjective matter. The diverse desiderata do, after all, enter into the scientific discussions concerning the choice, and even in the wake of revolutionary crises, a consensus usually emerges concerning the merits of competing new theories.

On this point, certain ideas advanced by Kuhn seem to me important and illuminating. Kuhn stresses that, in the end, the choice between competing theories is left to the specialists in the field. These specialists, he notes, have come to share broad standards for the appraisal of competing theories and for the choice among them. They acquire such shared modes of appraisal in the course of their professional training and their subsequent professional practice. But they do not learn them from books of rules that precisely specify the conditions under which a theory is to be given preference over its competitors. In interaction with their colleagues, in the course of their professional training and experience, scientists acquire fairly consonant dispositions to make preferential choices, but not dispositions to follow precisely specifiable rules of theoretical preference.

My characterization of scientific research as aiming at world pictures that are epistemically optimal at their time is thus indeed an oversimplified schematization. But I think my account is correct on the basic issue of my paper – namely, in pointing out that the goal of science has to be seen as epistemologically relativized, as consisting in the attainment of ever-changing world pictures, each of them epistemically optimal at its time.

As for the first part of my argument – namely, that science cannot be construed as a search for truth – it was based on purely logical considerations and thus is not affected at all by the vagueness of the desiderata.

Jointly, I hope, the two major parts of my argument suffice to indicate the basic soundness of a turn in methodology from a semantic-ontological to an epistemically relativized construal of the goal of science.

Probability

6   [5]* "Über den Gehalt von Wahrscheinlichkeitsaussagen," *Erkenntnis* 5 (1935), 228–60. Translation by Christoph Erlenkamp, revised by Olaf Helmer.

7   [14] "On the Logical Form of Probability-Statements," *Erkenntnis* 7 (1937), 154–60. [16] "Supplementary Remarks on the Form of Probability-Statements, Suggested by the Discussion. (Participants: Braithwaite, Clay, v. Dantzig, Ditchburn, Lindenbaum, Mannoury, Storer, Waismann.), *Erkenntnis* 7 (1937), 360–3.

8   [24] With P. Oppenheim, "A Definition of 'Degree of Confirmation,'" *Philosophy of Science* 12 (1945), 98–115.

Essay 6 is a translation of an article based on Hempel's Ph.D. dissertation. The essay argues for a view of probability as determined by empirical frequencies, while offering sympathetic but ultimately destructive criticism of the particular frequentist theories floated by Reichenbach and von Mises. The positive contribution of the article is a "finitization" of Reichenbach's approach.

The brief critical essay 7 also derives from the dissertation. It was read at the Fourth International Congress for the Unity of Science at Cambridge University in 1938. There, in a friendly spirit, Hempel identified an intractable difficulty that vitiates both von Mises's theory and Reichenbach's.

In the early 1940s, collaborating with Olaf Helmer and Paul Oppenheim, Hempel saw the project of logical empiricism as urgently needing a satisfactory account of how sentences can confirm or disconfirm other sentences. Unlike Carnap, the H_2O trio sought an explication of degree of confirmation in partially

---

* The numbers in brackets represent the publication number for each chapter here in this list. (See the section entitled "C. G. Hempel's Publications" beginning on p. 305.)

empirical terms. The high-water mark in their hopes for the project (1945) is marked by essay 8 and by a parallel publication by Helmer and Oppenheim.[1] Four decades later Hempel would conclude that, after all, the concept of confirmation belongs to pragmatics – not to syntax or semantics, not to logic.

Hempel's earlier nonquantitative account of the conditions under which observation sentences confirm general sentences[2] – for example, (1) "All ravens are black" – can be summarized as follows. Consider an observation report in which certain names appear – perhaps, in the ravens' case, just "*a*" and "*b*," where the report is this pair of sentences: "*a* is a black raven," "*b* is a non-black nonraven." This report is said to *directly* confirm (1) because it logically implies the restriction of (1) to the individuals mentioned in it, and it is said to *indirectly* confirm all other logical consequences of sentences it directly confirms. In the present example the restriction of (1) to "*a*" and "*b*" is the sentence (0): "If *a* is a raven, then *a* is black; and if *b* is a raven, then *b* is black." Since the report does logically imply (0), it directly confirms (1) and indirectly confirms (2) – "All young ravens are black."

This purely classificatory analysis of confirmation made no distinction between white shoes and black ravens as sources of confirmation for (1), even though black ravens do seem to confirm it and white shoes do not. But perhaps what is really going on is that both sorts of evidence do confirm (1), and the difference between them is a matter of degree. So Janina Lindenbaum-Hosiasson suggested.[3] Her idea is that while reports of black ravens and white shoes both confirm (1), the latter confirm it to a much lower degree, given our background information about rarity and homogeneity of ravens.

---

[1] "A Syntactical Definition of Probability and of Degree of Confirmation," *Journal of Symbolic Logic* 10 (1945), 25–60. A lovely account of H_2O by Nicholas Rescher appears in *Philosophy of Science* 64 (1997), 334–60.

[2] "A Purely Syntactical Definition of Confirmation," *Journal of Symbolic Logic* 8 (1943), 122–43. This early paper blossomed into Hempel's "Studies in the Logic of Confirmation," *Mind* 54 (1945), 1–26 and 97–121, which became chapter 1 of his *Aspects of Scientific Explanation* (New York: Free Press, 1965).

[3] *Journal of Symbolic Logic* 15 (1940), 136–41. She was killed by the Gestapo in 1942.

By 1945 Hempel seems to have taken her point. Anyway, essay 8 floats a quantitative account of degree of confirmation of hypotheses H by evidence-sentences E. If the language in which H and E are stated can classify members of the (finite) population into nonoverlapping, exhaustive categories, then a statistical distribution is an assignment to these categories of definite proportions in the population. Relative to different distributions, a sentence of the language may have different probabilities. Now the degree of confirmation of H given E is defined as the probability that H and E are both true divided by the probability that E is true – both probabilities being determined by a "maximum likelihood" distribution (i.e., a distribution making E as probable as any distribution can).

In Carnap's 1945 article entitled "On Inductive Logic,"[4] the role of the maximum likelihood distribution in this definition is played by the average of all possible distributions. Carnap's degrees of confirmation then satisfy all laws of probability, including one violated by the definition in essay 8 (i.e., the *multiplication principle*, according to which the degree of confirmation of *A-and-B* on E is the degree of confirmation of *A* on E times the degree of confirmation of *B* on *E-and-A*). Helmer, Hempel, and Oppenheim preferred their definition to Carnap's on the grounds of empiricism. Where Carnap assigned equal a priori probabilities to all possible distributions, they chose a distribution by reference to the evidence sentence E. They offered their definition as establishing fair betting odds on H given E, thus accounting for one aspect of confirmation. Other aspects – number and variety of instances reported in E – remained to be accounted for by some other concept, which would determine the amount it would be fair to stake in bets at those odds. (See Hempel's concluding remarks in Chapter 8 here.) But that view was hard to maintain in the face of Ramsey's and de Finetti's "Dutch book" arguments for the multiplication principle, arguments showing that "If anyone's mental condition violated [it], his choice would depend on the precise form in

---

[4] That is, in the same issue of *Philosophy of Science* 12 (1945), 72–97, that contained essay 8.

which the options were offered him, which would be absurd."[5]
This would be part of Hempel's later disenchantment with the
project of an inductive logic.

---

[5] Frank Ramsey, "Truth and Probability": see his *Philosophical Papers*, D. H.
Mellor, ed. (Cambridge, New York, and Melbourne: Cambridge University
Press, 1990), p. 78.

he backs away from inductive logic

Chapter 6

On the Content of
Probability Statements*

INTRODUCTION

The following inquiry attempts to contribute to the logical analysis of the concept of probability. Philosophical efforts to clarify
this concept are as old as its history, and the basic ideas of the
two main groups of theories which seek to determine its meaning
go back to the beginnings of this history. For a first orientation,
these may be contrasted as the "aprioristic" and "empiricist"
views of probability.

On the aprioristic views, probability theory rests on a priori
judgments of proportions in empirical or logical space. On the
empiricist views, probability theory concerns the relative frequencies with which certain results occur in statistical series of
empirical observations.

At present, these two views are no longer on an equal footing.
Instead, the progressive refinement of the logical analysis of
scientific concept formations has made it increasingly clear that
an adequate logico-descriptive theory of the probability concept
can be found only in the direction in which the empiricist interpretation seeks it.

A consistent realization of the empiricist view, however, leads

* From Carl G. Hempel, "Über den Gehalt von Wahrscheinlichkeitsaussagan,"
  Erkenntnis 5 (1935), 228–60. Reprinted with the kind permission of Kluwer Aca-
  demic Publishers. Translation by Christopher Erlenkamp, revised by Olaf
  Helmer.

to certain logical difficulties, which have not been completely overcome even in the latest articulations of the empirical-statistical probability theory. The present study examines these difficulties and develops a proposal for their removal on the basis of a "finitistic" turn of the statistical interpretation of the probability concept.

1. LOGICAL DIFFICULTIES OF THE STATISTICAL INTERPRETATION

The basic problem of the logical analysis of the probability concept – the problem in response to which the theory groups just mentioned differ from each other – is the question of the meaning of probability statements (i.e., of statements which attribute a certain probability to an event). Any answer to this question, which Reichenbach calls the "meaning problem of probability statements," also fixes the meaning of the probability concept in a certain way, for the logical function of a concept is exhaustively determined by its possible occurrences in sentences of a certain form, and the sense of these sentences fully fixes the meaning of the concept. Furthermore, any proposed solution to that meaning problem determines an answer to the question of the justification of probability statements (Reichenbach's *validity problem*, which he distinguishes from the meaning problem), for the sense of a sentence is determined by the totality of its testable consequences – and these also form the basis of any decision about its validity.

The empirical-statistical theories of probability, whose basic ideas agree in so many points that we shall also speak simply of "the" empiricist theory in what follows, start from the acknowledged fact that testing a statement that attributes a certain probability to a certain event does in principle always rest on ascertaining the relative frequency of that event in a long series of observations. Thus, for example, the statement that the probability of dying from a certain disease is 0.05 just says that in a sufficiently large series of afflictions of the relevant sort, 5% of the cases end fatally.

The more recent articulations of the statistical theory have tried to give a precise shape to this basic idea, a shape that would also be conveniently accessible to mathematical treatment. They start from the empirical finding that in statistical series of the sort to which probabilistic considerations are applied in empirical science, the fluctuations of an event's relative frequency become very small when the length of the series increases. On this, the statistical theory bases an "assumption" that lends itself particularly well to mathematical evaluation (i.e., that the relative frequency of a certain event converges toward a limit when such a statistical series is indefinitely extended). This limit is then called the event's probability in the statistical series.

As it stands, the resulting concept formation is vulnerable to a serious objection which H. Feigl[1] and H. Reichenbach[2] have developed most emphatically: According to the proposed interpretation of the probability concept, a probability statement has no empirical content at all. For on this interpretation, a probability statement merely specifies the limit of an event's relative frequency in an observation series that is thought of as infinitely extended. But obviously only finite statistical series are available for empirical tests, and from the value of the limit of a convergent numerical sequence, one cannot infer anything at all about the values of its terms within finite segments. Hence, a probability statement has no empirically testable consequences whatsoever. In other words, it has no empirical meaning, or, as we want to say in what follows, it lacks empirical content. Now, in empirical science one obviously subjects probability statements to an empirical testing procedure, so one does indeed attribute empirical content to them. In this essential respect the statistical theory thus fails to reflect the character of the probability concept correctly.

One may illustrate the objection this way. Given a probability statement that attributes probability w to the occurrence of a

[1] In his unpublished dissertation (Vienna, ca. 1928), and in his article "Wahrscheinlichkeit und Erfahrung," *Erkenntnis* 1 (1930), 249–59.

[2] "Kausalität und Wahrscheinlichkeit," *Erkenntnis* 1 (1930), 158–88; "Axiomatik der Wahrscheinlichkeitsrechnung," *Mathematische Zeitschrift* 34 (1931), 568–619; "Die logischen Grundlagen des Wahrscheinlichkeitsbegriffs," *Erkenntnis* 3 (1932), 401–25; and elsewhere.

certain event in a certain statistical series, consider the following possibilities:

1. More cases of the series in question are observed, and the relative frequency of the event in question is found to be approximately equal to w.
2. More cases are observed, but the event in question occurs with a frequency considerably different from w.

If one were to apply the above interpretation of the probability concept consistently, one would have to say that our probability statement is in no way better confirmed in the first situation than in the second. For the probability statement was said to specify only the *limit* of relative frequencies, and from an observation series however long, yet finite, one cannot infer anything at all about the correctness of the statement. Obviously, empirical science does not proceed on this principle.

The difficulty thus characterized occurs even in the most recent formulations of empiricist probability theory. In what follows, this will be shown by an examination of R. von Mises' and H. Reichenbach's theories.

2. ON THE LOGICAL FORM OF R. VON MISES' PROBABILITY THEORY

R. von Mises' theory, the essence of which[3] we must here assume to be known, is the first consistent realization of the limit-interpretation of the probability concept. It rests on the notion that probability statements are sentences of empirical science and that for this reason probability theory is to be developed as the mathematical theory of a certain branch of empirical science, just as formal geometry may be portrayed as a mathematical theory of a certain branch of empirical science (i.e., physical geometry).

It should be recalled that the logical analysis of the geometry problem led to a distinction between "formal" or "mathematical" geometries, on the one hand, and "physical" geometry, on the

---

[3] See especially his *Wahrscheinlichkeitsrechnung und ihre Anwendung* (Berlin and Leipzig: Deuticke, 1931).

other. A formal geometry is a system of sentences that are deducible from certain initial sentences, the axioms, by logic alone. Certain primitive concepts occur in the axioms, such as "point," "line," "plane," "congruence," "incidence," etc., which are not explicitly defined and are initially without content. The question of the validity of a formal geometry's theorems is to be answered by pointing out that each of them can be represented as a general implication with all axioms as its implicans and the theorem as its implicatum; in view of the deducibility just mentioned, any such implication is analytic, hence without empirical content. Given certain measurement conventions, physical geometry is, by contrast, a system of sentences with empirical content, a system that can be vaguely described as the physical theory of the positional relations of physical entities, in particular of rigid bodies. One can obtain this system from a suitably chosen formal geometry through a semantic interpretation of the primitive concepts proceeding by way of "coordinative definitions" – a method pursued in detail by Reichenbach.[4] The problem of developing a mathematical theory of "our world's" physical geometry may then be considered solved by specification of (I) an axiomatized formal geometry and (II) a system of coordinative definitions, such that the coordinative definitions turn the sentences of the formal geometry into those of the physical geometry.

In exactly the same way, von Mises wants us to understand the task of constructing probability theory as the mathematical theory of a certain branch of empirical science, and this is why he chooses the axiomatic method for constructing his theory. We shall sketch the basic ideas of this construction briefly, so that we may then compare its logical form with that of the axiomatic solution to the geometry problem.

The fundamental concept in von Mises' theory is that of a collective. As "the simplest collective," von Mises defines an infinite sequence of similar observations, such that the respective

---

[4] See for instance *Philosophie der Raum-Zeit-Lehre* (Berlin and Leipzig: de Gruyter, 1928); "Ziele und Wege der physikalischen Erkenntnis," in *Handbuch der Physik*, vol. 4 (Berlin: Springer, 1929), 1–80.

results of the observations can be represented by two signs – perhaps "0" and "1" – and the sequence meets the following two requirements:

Requirement 1. If among the first n observations, there are n_0 observations with the result "0" and n_1 observations with the result "1," then there exist the limits

$$\lim_{n\to\infty}\frac{n_0}{n}=w_0 \qquad \lim_{n\to\infty}\frac{n_1}{n}=w_1$$

Requirement 2. If an infinite partial sequence is constructed out of the total sequence by "place selection," then the same limits also exist within these partial sequences

$$\lim_{n\to\infty}\frac{n_0'}{n'}=w_0 \qquad \lim_{n\to\infty}\frac{n_1'}{n'}=w_1$$

Here, a "place selection" from an infinite observation sequence is the construction of a partial sequence on the basis of a rule which decides whether the nth observation is a member of the partial sequence, and it does so independently of the result of this nth observation. At most, the results of the preceding observations are taken into account.

The concept of probability is now defined with reference to collectives alone; that is, the limit w_0 (or w_1) introduced above is called the probability with which the feature "0" (or "1") occurs in the collective under consideration. (In the present context, we may disregard the general concept of a collective, which is explicated in terms of the one just defined.)

The conditions characterizing a collective which we have just presented are the axioms in von Mises' probability theory. (He calls the second requirement the "irregularity axiom.") This theory is nothing but the purely mathematical or logico-deductive theory of the limits of relative frequencies in sequences that have the character of a collective. As von Mises emphasizes, the probability calculus has only this function: From given limits of relative frequencies in certain initial collectives, it establishes the limits of relative frequencies in certain new collectives that are "derived" from the original collectives in accordance with certain specifiable methods. Following the distinction we introduced in connection with the geometry problem, we shall first look at (I) the logical form of the mathematical theory, and then ask about (II) the coordinative definitions that turn the mathematical theory into a branch of empirical science.

Regarding (I). To begin with, the following difficulty arises in the formal construction of von Mises' theory. The irregularity axiom leads to the conclusion that numerical sequences with the character of a collective cannot be given intensionally (i.e., by means of a general rule of construction for its members);[5] but an infinite mathematical sequence can in principle be determined only by its construction rule. Hence von Mises' concept of a collective is not a proper mathematical concept. The intramathematical difficulties that arise in this context have already been thoroughly examined several times.[6] For this reason we shall not discuss them any further.[7]

All the same, von Mises accepts the irregularity axiom as part of the foundations of his theory in order to obtain an explicit definition of the concept of a collective from which all theorems of the probability calculus become mathematically derivable. With this objective, however, the principle of a proper axiomatization of the probability calculus is abandoned. For let us suppose that the difficulties surrounding the irregularity

---

[5] See, for example, R. v. Mises, "Grundlagen der Wahrscheinlichkeitsrechnung," *Mathematische Zeitschrift* 5 (1919), 52–99.

[6] Especially by K. Dörge, "Zu der von R. v. Mises gegebenen Begründung der Wahrscheinlichkeitsrechnung," *Mathematische Zeitschrift* 32 (1930), 232–58; E. Kamke, "Über neuere Begründungen der Wahrscheinlichkeitsrechnung," *Jahresbericht der Deutschen Mathematische Vereinigung* 42 (1933), 14–27, 182; E. Kamke, *Einführung in die Wahrscheinlichkeitstheorie* (Leipzig: Hirzel, 1932); H. Reichenbach, "Axiomatik der Wahrscheinlichkeitsrechnung," *Mathematische Zeitschrift* 34 (1932), 568–619; F. Waismann, "Logische Analyse des Wahrscheinlichkeitsbegriffs," *Erkenntnis* 1 (1930), 228–48.

[7] *Editor's Note:* Real progress toward satisfactory formulations of the regularity requirement came only in the 1960s, notably in the work of Per Martin-Löf. For references, see R. Jeffrey, *Probability and the Art of Judgment* (Cambridge and New York: Cambridge University Press, 1992), chapter 11.

requirement were overcome – that perhaps a suitable substitute requirement had been introduced. Explicit definitions of the concepts "collective" and "probability" would then stand at the top of the theory, and all theorems of the theory would then be deducible from these definitions by purely logical means. Since by assumption the initial definitions contained only logico-mathematical concepts, no extralogical primitive sign would occur in the entire system, and it would therefore be impossible to give the system any semantic interpretation. Such a theory could in principle not serve as the formal framework of an empirical theory. (This consideration applies also to those articulations of the statistical theory that, like Kamke's,[8] base the definition of the probability concept on the concept of a limit in mathematical analysis and dispense with a precondition corresponding to the irregularity axiom.) Conversely, one may safely say of von Mises' theory that its claim to be empirically interpretable is expressed by the very fact that the irregularity requirement cannot be represented by means of purely logico-mathematical concepts. But the irregularity requirement does not have the form of a genuine axiom, nor does it even suggest an approach to its axiomatization.

Regarding (II). As to the question concerning coordinative definitions, it should be first noted that von Mises himself does not explicitly introduce the distinction between setting up an axiomatic system and specifying an empirical interpretation. Nevertheless, in the fundamental definitions of his theory and in the added explanatory notes, one finds detailed clues that allow quite a precise characterization of empirical collectives as he conceives of them.

Thus, the following coordinative definition should do justice to von Mises' view. By an empirical collective one is to understand an indefinitely extensible series of like empirical observation data which satisfies the two conditions expressed in the two axioms of the probability calculus. By this stipulation, the concept of probability receives its physical interpretation as well.

[8] See footnote 6.

But this definition of the concept of an empirical collective is vulnerable to the objection we developed earlier. According to von Mises, all probability statements refer to the limits of relative frequencies in collectives; therefore, they are in principle not testable by finite means, the only means empirical science has at hand. Hence, they lack empirical content.

In response to such objections, von Mises has maintained that the application of the concept of an infinite sequence to problems of physics is an idealization, one that does not differ in kind from the idealization one makes when applying geometrical concepts to physics. The claim that a certain physical entity has the shape of a sphere, for example, presupposes infinitely many measurements; in actual fact, one decides already after a sufficiently large *finite* number of measurements to make an *idealization*; and it is for reasons of convenience that one describes the finite totality with the aid of the concept "sphere."

In the same way, von Mises continues, one introduces an idealization on the basis of only finitely many observations when one applies the probability calculus to questions of physics. "For *practical* reasons one uses the infinite theory; although the experimental material is drawn from finite segments, one calculates as if it were infinite . . ."[9] Kamke inclines to the same view.

Yet it can be shown that this comparison between probability theory and geometry is incorrect. Let us first consider the statement "This body is a homogeneous sphere." It obviously determines an indefinitely extensible set of empirically testable consequences. For instance, one can specify precisely how to find a point inside the body such that two very thin channels drilled toward it from any two points on the body's surface will be of equal length. Similarly, on a horizontal plane the body will be in neutral equilibrium. And again, from an overflow receptacle it will displace a volume of liquid which stands in a certain arithmetical relation to the length of its diameter (i.e., a line segment to be measured in accordance with certain instructions). And so on.

[9] v. Mises et al., "Diskussion über Wahrscheinlichkeit," *Erkenntnis* 1 (1930), 260–85.

Testing these consequences now permits empirical checking of our initial statement. Of course, since the set of consequences is indefinitely extensible, this can never lead to a complete and final establishment of the truth of our initial statement, but only to a gradually improving confirmation – or to a refutation, as when we encounter empirical findings that are incompatible with the initial sentence (e.g., when a result of the liquid-displacement experiment deviates from the relevant prediction). This consideration will be sharpened in a later section of this essay. Following the usual terminology, we now want to call the claim that a certain physical body is a sphere a *hypothesis*. It is not completely and finally decidable, but it is "checkable" in the sense that it leads to the specification of any number of testable consequences.

The "idealizing" claim that a given set of observational data determines a collective is of a fundamentally different character. For not a single testable sentence is derivable from this claim, since from the existence – or even from the value – of the limit of an infinite sequence, no conclusions can be drawn about the values of the terms in any finite segment. But since only such finite segments are accessible to empirical checking, no empirical discovery can in principle be described as in accordance or not in accordance with the claim – indeed, one can apparently not even speak of an *approximate* test. In short, in contrast to the case of a genuine hypothesis, here we are dealing not with an infinite totality of statements but instead with a pseudostatement about an infinite totality.

What is actually ascertained in an empirical confirmation of a probability statement is, as it were, much more than the existence of the limit in question. It is the fact that when the observation series is extended, the relative frequencies to be examined are already almost constant in finite, conveniently accessible domains. Later on, this idea will be taken up again.

Nor is the transfinite characterization of the empirical collective justifiable by pointing out that the inference from finitely, many observed data to the existence or nonexistence of a collective with its characteristic transfinite properties is simply an inductive inference, of the sort constantly made in empirical science. For empirical inductive inferences always lead to results that are "checkable" in the sense mentioned above; and from this it follows right away that the uncheckable claim that a given finite set of observation data determines a collective cannot be the result of an inductive inference.

A further consequence of the above considerations is worth pointing out at this point. As already mentioned, the basic approaches of the statistical theory generally rely on the "assumption" that a curious law holds in our world, one which is revealed in the "experiential fact that in certain series of phenomena the relative frequency of an event's occurrence approaches a certain limit when the observation is continued indefinitely."[10] It is now evident that because of its transfinite character, this "assumption" – often called the "Law of Large Numbers" – lacks empirical content and that it must, therefore, not be treated in parallel to the genuine laws of empirical science.

Finally, the considerations under (I) reveal that the validity of the theorems of probability theory is self-evident for transfinitely interpreted collectives, since by definition such collectives satisfy those conditions from which all theorems of the theory follow analytically. So the empirical content of the theory is put into the coordinative definition and not into the axioms. To be certain on the basis of the coordinative definition that a statistical series of empirical data determines a collective – that one may, therefore, apply probability theory to that series – one has to know at least as much about the series as the theorems of the probability calculus will subsequently allow one to say about that series. Now, of course, the validity problem no longer arises for the probability calculus, but the theory does not enable us to advance non-analytic theorems, either. The discussion of the problem of empirical interpretation thus shows the difficulty already mentioned in (I) from a new angle.

One should bear in mind that the coordinative definitions of geometry are of a completely different character. They do not single out the entities that are to be called "physical points,"

---

[10] v. Mises, *Wahrscheinlichkeit, Statistik und Wahrheit* (Vienna: Julins Springer, 1928), p. 91.

"physical lines," etc., by the requirement that these entities must satisfy the semantically interpreted axioms of formal geometry. Instead, they single them out without any reference to the axioms whatsoever. (They proceed in the sense of this schema: points = tiny bodies, lines = light rays, etc.) In this way the coordinative definitions turn the theorems of pure geometry into sentences of physics, not all of which are analytic.

As a result of the above, a logically satisfactory articulation of the empiricist theory of probability faces two main tasks:

I.   the task of formalizing the physical probability theory in an axiom system that satisfies the conditions developed under (I) above;

II. the task of specifying a finite interpretation of that axiom system's primitive concepts, which does not rely on the condition that the empirical entities to be characterized have the properties formulated in the axioms.

3. ON H. REICHENBACH'S AXIOM SYSTEM OF THE PROBABILITY CALCULUS

Lately, the logical analysis of the probability concept has been considerably advanced due to H. Reichenbach's investigations.[11] We will try to characterize their importance for the issues sketched at the end of the previous section. Again, we must assume that the content of these investigations is essentially known.

Regarding (I). Reichenbach develops an axiomatization of the probability calculus which satisfies the above-mentioned conditions. In this axiomatization, each probability statement is represented in the following form:

$$(i)(x_i \in O \xrightarrow[p]{} y_i \in P)$$

[11] See, in particular, "Axiomatik der Wahrscheinlichkeitsrechung," *Mathematische Zeitschrit* 34 (1932), 568–619, and "Die logischen Grundlagen des Wahrscheinlichkeitsbegriffs," *Erkenntnis* 3 (1932), 401–25; cited in what follows as "Axiomatik" and "log. Grundlagen."

which may be abbreviated as

$$O \xrightarrow[p]{} P$$

In addition to sign combinations of this form, the axioms of the "formal probability calculus" – as Reichenbach says by analogy with the expression "formal geometry" – contain the logical primitive concepts of the propositional calculus and of the lower functional calculus; one derives the theorems of the formal probability calculus in accordance with the inference rules of these calculi.

So Reichenbach's axiom system of the probability calculus does indeed contain nonlogical primitive signs (e.g., the sign $\xrightarrow[p]{}$ of "probability implication of degree p") and also lowercase Latin letters with subscripts (e.g., x_i, y_i [i = 1, 2, 3, . . .]) to denote the members of two denumerably infinite sets that are mapped one-to-one onto each other – and also uppercase Latin letters (such as O, P) to denote variable classes that can, on account of their logical type, contain the x_i or y_i as members.

At bottom, Reichenbach's representation of the form of a probability statement is a logistically more precise rendering of von Mises' basic idea that probability is a relational concept, which can be defined only by reference to a collective. But the virtue of Reichenbach's theory that is of special interest here lies in the fact that in the spirit of the axiomatic method, probability theory gets constructed purely formally, without any semantic interpretation of the nonlogical primitive concepts. In particular, it is not presupposed that probability implications are limit statements. Rather, it is shown that the limit reading of the statistical concept of probability can be regarded as an *arithmetical* interpretation of the formal probability implication. In virtue of this interpretation, each formula of the formal probability calculus turns into an analytic arithmetical sentence, and in this way one obtains an "arithmetical model" of the formal probability calculus. This model contains those and only those sentences of the mathematical-statistical probability theory which are derivable solely from the precondition that the sequences under consideration converge (requirement 1 of von Mises' theory).

Regarding (II). To arrive at an *empirical* interpretation of his axiom system, Reichenbach (in "Axiomatik") uses a coordinative definition. We may put the basic idea of this coordinative definition as follows: by x_i, y_i ($i = 1, 2, 3, \ldots$), one is to understand the members of two indefinitely extensible sets of empirical observation data, where these members are assigned to each other in pairs. By O, P, one is to understand two classes of which the x_i or y_i, respectively, can be members. Finally, one is to interpret the formula

$$(i)(x_i \in O \xrightarrow[p]{} y_i \in P)$$

by the following statement: "The relative frequency of those cases in the data series in which $x_i \in O$ and $y_i \in P$ holds, relative to the number of cases in which $x_i \in O$ holds, converges toward the limit p when the number of cases increases."

However, if one wants to interpret the sentences of the formal probability calculus by means of this coordinative definition, one again faces the difficulty discussed above. Since the interpretation rule is of transfinite character, one cannot decide for any empirical observation material whether it satisfies the conditions of the coordinative definition or not. Reichenbach (in "Axiomatik") regards this difficulty as insurmountable if one acknowledges as meaningful only those claims that can in principle be decided definitely as true or false. For this reason, he makes the transition from "strict logic" to a "probabilistic logic." Within probabilistic logic, probability statements – though not decidable as true – turn out to be still decidable as probable. Reichenbach regards this as reasonable and sufficient for probabilistic logic.

Thus armed, Reichenbach assails the problem posed by the transfinite character of his coordinative definition, and he does so on the basis of the following consideration. The decision whether or not a given series of physical observations has a limit cannot be a matter of empirical discovery, but it must be a matter of axiomatic stipulation. To secure the applicability of this coordinative definition, it is therefore necessary to introduce a special axiom. This reads as follows.

Axiom of induction: If a rule of physics determines an indefinitely extensible sequence of frequency quantities h_i (i.e., quantities that either are relative frequencies or result from such by certain arithmetical operations) of which $h_1 \ldots h_n$ have already been ascertained, and if s_1 and s_2 are numbers such that from a term h_m ($1 \leq m \leq n$) onward, all h_i ($m \leq i \leq n$) lie between the bounds s_1 and s_2, then there is a probability of w that the sequence tends to a limit between s_1 and s_2; w increases toward 1 if $n - m$ goes toward infinity while all h_i ($m \leq i \leq n$) stay between these bounds. ("Axiomatik," p. 614)

With that, we have outlined the main features of the probability theory that Reichenbach develops in "Axiomatik." In his most recent publications, Reichenbach has modified his foundation of the concept of probability. Without departing from the transfinite form of empirical probability statements, he avoids introducing a special induction axiom to secure the applicability of his interpretation rule. We will return to this more recent view at the end of this section.

1. Let us set aside the question of applicability for the time being. Let us suppose that it is possible to decide whether two series of empirical observation data x_i, y_i, when extended infinitely, satisfy the limit condition expressed in the above coordinative definition. But then the objection to von Mises' theory developed above still stands. That is, if we chose this interpretation rule, before we could even apply the probability calculus to some empirical material, we would have to know as much about that material as all formulae of the calculus would later permit us to say about it; for we would need to know that the sequences under consideration have limits. And as Reichenbach himself shows when presenting the arithmetical model of the formal probability calculus, all theorems of that calculus are analytic sentences for convergent sequences – apparently it does not matter whether their members are purely arithmetical or empirical data. Therefore, the theorems of the probability calculus that refer to some empirical material according to Reichenbach's coordinative definition turn out to be analytic.

2. Now, to put it briefly, the induction axiom has the function

of undoing, as it were, the relocation of the theory's empirical content into the coordinative definition. Indeed, the induction axiom is meant to allow us to apply the formulae of the probability calculus already to finite series of empirical frequency quantities. Yet this method of securing the applicability of the probability calculus is vulnerable to certain objections.

Reichenbach reduces the validity problem of the applied probability calculus to the problem of the validity of the induction axiom ("Axiomatik," p. 618); the latter is thus treated like a law of nature. But since the induction axiom, because of its transfinite character, lacks empirical content, there is no validity problem for the induction axiom! (It is apparently in view of this consideration that Reichenbach talks explicitly of an *axiom* and that he explains that it is introduced by an "axiomatic requirement going beyond what can be ascertained." But this remark does not remove the difficulty. Instead, this remark makes it even clearer that there cannot be a validity problem for the induction axiom.)

Furthermore it is easily seen that for *every* sequence of frequency quantities of the sort the induction axiom mentions, one can specify three natural numbers m, s_1, s_2 such that the conditions of the axiom are satisfied. It follows that, according to the axiom, the formulae of the probability calculus are applicable with a certain probability to *any arbitrary* finite series of frequency numbers. The induction axiom in its above form seems to say "too much" in one respect, and "too little" in another respect. It seems to say too much in that it does not distinguish in any way between the usual "empirical collectives" and arbitrarily arranged sequences of physical frequency quantities. It seems to say too little in that it is incalculably weakened by the indeterminate clause "with a certain probability."

For all these reasons it seems unacceptable to introduce the induction axiom as something like a law of nature. Rather, the axiom is perhaps more appropriately characterized as expressing a stipulation – one that does not concern the quasi-empirical convergence of physical sequences when they are "indefinitely extended" but that concerns instead the semantic interpretation of the formulae of the form $O \xrightarrow{\quad} P$. Instead of interposing the

induction axiom and saying, "Finite series of empirical frequency numbers have, with a certain probability, a limit, and for
sequences with a limit the formulae of the probability calculus
are analytic," one should stipulate, in the sense of stating a coordinative definition, "The formulae of the probability calculus are
applicable to finite series of empirical frequency quantities." To
me, this seems to suit the logical situation better. By way of such
a stipulation – which, of course, calls for a more careful formulation – the probability calculus is presented as a formal theory
of a branch of empirical science, and the problems surrounding
the induction axiom are avoided as well.

3. The objections developed in (2) do not apply to the
somewhat modified form that Reichenbach has given to his
theory in his most recent publications. In "log. Grundlagen" the
résumé of his present views does indicate that Reichenbach still
adheres to the axiom system and to the transfinite coordinative
definition (hence, the considerations on this score, set forth in
[1], remain in force). On the other hand, the applicability of the
coordinative definition, which presupposes convergent empirical
sequences, is no longer assured by means of an axiom. Instead,
Reichenbach bases the application of the probability calculus
on a "theory of posits." As to the details of this concept, one
should consult the article mentioned; here it should suffice to
quote only the passage that is crucial to our present context:
"When we have observed a certain frequency h^n in the sequence's
finite segment at hand, we posit that the sequence tends to a limit
at h^n (more correctly: within $h^n \pm \delta$). We posit this; we do not want
to say that this is true; we posit it merely in the sense in which a
gambler bets on the horse that he considers the fastest" ("log.
Grundlagen," p. 414). Reichenbach also seeks to *justify* this
scientific procedure – he calls it "approximative positing" –
which has been only *described* so far. The basic ideas of his argument may be summarized as follows. If a series of empirical data
under consideration is convergent at all, then the procedure of
"approximative positing" leads certainly to a gradual approximation of the value of its limit; if, however, we are dealing with
a divergent sequence, no procedure at all can succeed. It is true
that we can never know of an empirical sequence whether it is

convergent, but we may well say that if anything is to be achieved at all, the procedure of approximative positing will get us there.

This approach seems to revert to an idea on which the introduction of the induction axiom was based. The validity of the induction axiom was meant to support the theorems of the applied probability calculus; in the same way, the presupposition that the empirical sequences have a limit is here said to license approximative positing. The difference is just that the validity of this presupposition is not postulated, but instead is presented as problematic. The gist of the objections developed in (2) applies *mutatis mutandis* to this consideration. At any rate, these objections should remain in force as long as the transfinite approach to empirical data series is treated as a claim or as a presupposition with empirical content.

To put it positively: Again, a finitistic coordinative definition is most likely manifested in the idea that "empirical probabilities are to be determined from finite data series through approximative positing; approximative positing leads to correct results if the data series determine convergent sequences; in this case, the theorems of the probability calculus hold for the numbers found by the finitistic positing-procedure." And this coordinative definition essentially says that the formulae of the probability calculus are applicable to finite series of empirical frequency quantities. Such an approach also makes the application of the probability calculus independent of Reichenbach's justification of approximative positing with its transfinite-empirical presuppositions. Just as the coordinative definitions that turn formal geometry into applied geometry get no special "justification," so too does it hardly seem required – or even admissible – to justify or substantiate with additional considerations the coordinative definitions that are implicitly introduced with the concept of approximative positing.

In what follows, the possibility of a finitistic interpretation of the empirical-statistical probability concept, which has already been hinted at several times above, will be examined more closely.

4. ON THE "FINITISTIC" INTERPRETATION OF THE EMPIRICAL PROBABILITY CONCEPT

A finitistic interpretation of the concept of probability is not only suggested by an inspection of the logical difficulties surrounding the transfinite interpretation that we have discussed so far. A finitistic interpretation arises also out of a purely descriptive analysis of the methods by which probability statements are advanced or tested in empirical science. This will now be explained in greater detail. (When doing so, we will frequently limit ourselves to examining probabilistic considerations in physics. In principle, they do not differ at all from those of other branches of empirical science, but because of the precision of concept formation in physics, they are perhaps most accessible to a logical analysis.)

a. It is clear that the test of an empirical probability statement is always based on some finite material. In particular, the laws of the probability calculus are also applied to such observation series as are not even "in principle" indefinitely extensible, as the application of the limit concept of probability would require. Consider, for instance, the field of social statistics, and further-more all applications of the probability calculus in physics more narrowly conceived, insofar as they are based on group averages. In these cases, "indefinite extensibility" of the statistical material would require that there be "infinitely many" similar things or events in the world. Thus, when the probability calculus is applied to electron statistics, there would have to be infinitely many electrons. Irrespective of the fact that the meaning of such a hypothesis would still remain to be thoroughly clarified, one may at any rate say that the application of probabilistic methods in physics is not made dependent on whether or not such pre-suppositions are considered reasonable and valid. Something analogous holds for cases in which the probability calculus is applied to temporal averages.

b. The finitistic character of empirical probability statements emerges very clearly in contexts where statistically interpreted probability values are brought into relation with values of other quantities. This happens, for instance, when in the framework of Bohr's theory the intensity proportions of spectral lines are identified with the proportions of the corresponding transition probabilities. What this approach amounts to is that the probability of a certain transition is replaced by that transition's relative frequency in a finite set of quantum leaps, since it is certain that only finitely many quantum leaps occur in a finite observation time. Without such a finitization, the concept of probability would obviously not be suited for generating this checkable statement about intensity proportions. The case is similar when from the decay probability of a radioactive substance, one computes in a well-known manner how many of n atoms of this substance are still in existence at the end of time Δt.

These considerations, too, make it seem appropriate to frame the semantic interpretation of probability statements in a finite way. And indeed, in the discussion of the problem of probability, finitistic tendencies have already emerged several times. An instructive example of such endeavors lies in the question that O. Neurath posed in 1929 at the Prague Conference for Epistemology of the Exact Natural Sciences. He asked "whether the concept of the infinite might not be avoided by erecting the edifice of the mathematical probability calculus, not on the concept of the collective, but on the concept of a *finite* set of elements instead" (*Erkenntnis* 1 [1930], "Diskussion," p. 277).[12] R. von Mises replied at that time,

> ... that would be very difficult and unperspicuous, and one would not attain a single simple theorem. The first step made by an insurance company in its calculations already presupposes an infinite population. Accordingly, matters stand as follows: For *practical* reasons one uses the infinite theory;

---

[12] Meanwhile, the main features of the construction of a probability theory as a mathematical theory of finite collectives have been worked out by Hans Blume in the following two works: (1) *Zur axiomatischen Grundlegung der Wahrscheinlichkeitsrechnung* (Ph.D. dissertation, University of Münster, 1933); (2) "Mathematische Begründung und Entwicklung einer Wahrscheinlichkeitsrechnung mit finiten Kollektiven," *Zeitschrift für Physik* 92 (1934), 232–52.

although the experimental material is drawn from finite segments, one computes as if it were infinite, and in all operations one uses the computational rules that rest on the assumption of infinite collectives, because these rules are shorter, more simple, and more perspicuous. (*Erkenntnis* 1 [1930], "Diskussion," p. 279)

In this discussion, however, one does not clearly distinguish between the logical structure of the sentences of a formalized theory and the logical structure of the sentences correlated with the former by the interpretation rule. As the preceding reflections show, removing the logical difficulties of the probability concept does by no means call for finitizing the treatment of the calculus aspects of probability statements. It is essential only that the empirical interpretation of probability statements not be given a transfinite character. It is not the formulae of the calculus that should be finitized, but those sentences by which the empirical content of empirical probability statements gets represented.

The insight that Reichenbach's axiom system makes possible is invaluable for clarifying this logical situation. The limit interpretation of the probability concept characterizes only one of the models of the formal probability calculus, but the limit interpretation is by no means necessary for the construction of the formal theory itself. The failure to appreciate this fact has most likely contributed to the belief that the application of the probability calculus should also be limited to infinite sequences.

But there are still other considerations that have fostered adherence to a transfinite interpretation of empirical probability statements. The idea of viewing probability statements as statements about relative frequencies in finite data series (a) seems to lead to downright absurdities in some cases, and (b) does not in any case seem to enable us to represent the meaning of empirical probability statements.

Indeed, a finitistic interpretation of the statement, say, "The probability of throwing a five with this die is $1/6$" would, in a first approximation, read as follows: "In every series of n throws of this die, where n is an arbitrary finite number, $n/6$ fives are thrown." This is now (a) absurd for such values of n that are not integral multiples of 6. (b) But regardless of this, the probability statement in question just does not have the meaning of specifying the relative frequencies of fives as exactly 1/6. When certain deviations occur, one does not immediately call the probability statement false; on the contrary, even sizable deviations are regularly expected.

Now it is true that these considerations can by no means justify the transfinite interpretation of the probability concept, since they obviously leave all those objections completely untouched that we have brought forth against it above. Yet these considerations may well serve as suggestions for a suitable qualification of the finitistic reading of probability statements that we just adumbrated.

As we will show shortly, difficulty (a) can be easily avoided with a suitable modification of the finitistic interpretation. Objection (b), on the other hand, does not seem at all to reveal a logical difficulty that attaches to just the finitistic interpretation of probability statements and that proves its untenability. Rather, the objection merely seems to shed light, via the example of probability statements, on a logical situation that holds in principle for every sentence of empirical science.

For objection (b) presupposes the "classical" view of hypotheses to which we alluded earlier. According to this view, a hypothesis is a universal sentence (more precisely, a universal implication); it comprises infinitely many individual or special instances, not all of which can be tested, and it can, therefore, itself not be completely and finally confirmed. On the other hand, a hypothesis can be finally refuted; the proof that a single one of its special instances is an empirically false sentence suffices for this. Applied to our case, this means: According to the proposed finitistic interpretation, the statement about the probability of throwing a five is a hypothesis that holds universally for every value of n. The individual instances of the hypothesis are obtained by substituting arbitrary integral values for n. Thus, one of these individual instances reads, "Among 24 casts with this die, exactly 4 fives are thrown." If the finitistic interpretation were correct, then our initial probability statement would already have to count as refuted when, in a test, this single consequence alone turned out to be false. But in actual fact, the probability statement is by no means necessarily called false in such a case. For this reason, the finitistic interpretation cannot capture the meaning of empirical probability statements.

It can be shown, however, that an empirical sentence of whatever logical form is in principle neither finally and completely refutable nor confirmable. The "classical" view of hypotheses is thereby seen to be too narrow, and objection (b), which rests on it, no longer applies. This will now be explained in greater detail.

The following consideration leads to a formal characterization of the procedure by which empirical science tests its sentences.[13] Empirical science is a system of sentences which are constructed from certain basic components according to certain "syntactical" rules and which stand in certain inferential relations to each other. This system of sentences is set up as modifiable; thereby, our formal approach takes account of the changeability of the body of empirical knowledge which is due to progress in research. That is, gaining a new piece of empirical knowledge is manifested by the incorporation of a "new" sentence (or, of course, several "new" sentences) into the system, which will perhaps require deleting certain "old" sentences. In this connection we call a sentence "old" or "new" depending on whether or not it is a logical consequence of the sentences that are already accepted.

How does one decide, then, whether or not a new sentence proposed for incorporation – irrespective of its logical form, we want to call such a sentence a "hypothesis," too – gets incorporated into the body of empirical sentences? One usually says, in the customary "semantic" (Carnap) manner of speaking, that to each sentence there belong certain verification conditions, which allow us to establish the truth or falsity of the sentence in question by suitable observations or experiments. (There

---

[13] Following Carnap, "Über Protokollsätze," *Erkenntnis* 3 (1932), 215–28; and Neurath, "Protokollsätze," *Erkenntnis* 3 (1932), 204–14. Compare also Hempel, "On the Logical Positivists' Theory of Truth," *Analysis* 2 (1935), 49–59 (reprinted in Chapter 1 here).

are no differences of principle between the "observational" and "experimental" methods; for this reason, we need not always explicitly carry out this terminological distinction in what follows.) The test procedure is thus based on suitable experimental checks, and it makes the truth of the sentence dependent on the outcome of these experiments. It is of no epistemological significance here whether the experimental checks are made before or after the hypothesis to be tested has been advanced. In the practice of gaining knowledge, it is, of course, possible that there is an important difference here. The first case may be a matter of "advancing a hypothesis on the basis of certain empirical data," whereas the second may be a matter of "subsequently testing" a hypothesis – one otherwise gained, or one set up tentatively on the basis of a conjecture – by empirical material procured especially for this purpose. The logical structure of the checking procedure, however, is the same in both cases. Each time, the statement to be tested, the "hypothesis," is confronted with the total body of accepted empirical sentences.

This confrontation conforms to the following scheme. From the system S of accepted empirical sentences, suitable universal or singular sentences $p_1, p_2, \ldots, p_k$ are adjoined to the hypothesis h; from the set of sentences $\{h, p_1, p_2, \ldots, p_k\}$ thus formed, certain conclusions $f_1, f_2, \ldots, f_l$ are drawn, and these conclusions are then compared with suitable further sentences $g_1, g_2, \ldots, g_m$ from S.

> Example. Suppose that h means, "The density of mercury at room temperature is 13.5." Suppose further that one tests h by weighing a receptacle filled with mercury. Then the system of sentences p_x contains the definition of the concept "density" and, moreover, perhaps the specifications "The receptacle has a volume of $50 \, \text{cm}^3$" and "The empty receptacle weighs $30 \, \text{g}$." From these sentences, together with the hypothesis h, one can draw the conclusion f, "At room temperature, the receptacle filled with mercury weighs $30 + 13.5 \cdot 50 = 705 \, \text{g}$." Now, a "suitable" sentence of S, with which f can be compared, is a sentence that states the weight of the full receptacle as established by weighing it. It

reads, perhaps: g, "The receptacle filled with mercury weighed 703 g."

We want to call those sentences g_m from S by which a hypothesis is checked "control sentences"; furthermore, those sentences f_l that are derived from the hypothesis and appropriate further premises are to be called "control consequences" of the hypothesis.

It should be noted that any sentence from S can in principle function as a control sentence. In science, control sentences – and hence control consequences as well – are not subject to restricting conditions. In particular, they are neither limited to the class of the formally most simple sentences of the language of physics, nor to the class of the so-called observation statements. (It is useful to compare Carnap's detailed explications in this context; see Carnap, "Über Protokollsätze," p. 222 ff.; fn. 13 in the present essay.) The syntactic observations sketched here characterize for Carnap one among several possible forms of language – that is, the one he calls the "second form of language approach B" (i.e., protocol sentences without restrictions in the system language). This form of language is the only one we have considered here, since it represents the structure of present-day science most precisely. Carnap is most likely to share this view.

These last considerations show that the basic idea of the classical view – that a hypothesis can, in principle, not be finally confirmed – must be extended to all empirical sentences. One can indeed specify any number of control possibilities for every empirical sentence. (In the above example dealing with the density of mercury, there are still numerous other methods of measurement, but one can also indefinitely modify the method we took above as basic by changing the constants that enter into the measurement, such as the volume or weight of the receptacle, etc. Each time, one arrives at other control consequences.) In other words, one can extend the system S in any number of different ways by introducing new sentences (to speak "semantically": by realizing ever new experimental conditions for testing

113

h) such that the set of control consequences of *h* gets extended as well. *In this sense* we want to say that *a hypothesis has "indefinitely many" control consequences.*

The classical assumption that it is possible to refute empirical sentences with finality, on the other hand, cannot be sustained. The idea on which this assumption rests can now be phrased as follows. A hypothesis *h* is not incorporated into *S* if there is at least one control consequence *f* of *h* that does not agree with the corresponding control sentence *g*. Yet, this view does not conform to the test procedure actually practiced in science, as Reichenbach in particular has stressed several times. The statement about the density of mercury we considered a little while ago, for example, does not necessarily count as refuted if, as we assumed above, the "empirical findings" (control sentence) do not agree exactly with the "prediction" (control consequence). In principle, the same holds for every other case. Control consequences that have the form of numerical specifications (perhaps of length, temperature, etc.) are particularly instructive examples. In these cases one is content when the respective quantity to be measured falls within a sufficiently narrow interval around the value specified by the control sentence. And it may be said that a hypothesis is not only not necessarily given up when one of the control experiments leads to a deviating result, but also that in many cases, especially of metric specifications, a hypothesis gets "confirmed" exclusively on the basis of measurements that show small deviations from the results that the hypothesis leads one to expect. A typical case of this sort arises when a curve, one intended to represent an empirical correlation, gets constructed from observation points, not one of which lies on the curve itself.

If a hypothesis is incorporated into the system *S* despite deviations of the sort indicated, then the system's body of sentences must be suitably revised so that consistency is secured. Here, in most cases, some of the sentences p_x or some of the sentence g_m [is] discarded from the system *S* for this end. (In the simplest case this means, intuitively, that one assumes errors in the experimental or observational procedure.) Consistency can be secured by still other revisions, which, however, are not going to be discussed here.

One could now try to capture the basic idea of the classical view about the refutability of a hypothesis in a form that is better adapted to scientific practice. A hypothesis counts as refuted if at least one control experiment yields a result that deviates by more than a fixed maximum from the result the hypothesis leads one to expect. In physics, hypotheses of metric form are indeed often supplemented with an explicit remark about maximal deviation. Sometimes, however, a hypothesis is still not given up even if that interval is occasionally overstepped. So the classical view about a hypothesis's final refutability by a single counter-example is incorrect even for those hypotheses that come with a specified interval. The question concerning the criteria of falsification arises, as it were, on a higher level, all over again.

As a last recourse, one might try to uphold the classical view in the form of the following more general claim. Acceptance of a hypothesis does not require that every single measurement result agree with the value yielded by the hypothesis, but only that a suitably defined average of the results of a series of like measurements agree with it. At times, however, a hypothesis is retained even if there is no exact agreement of this sort. This is quite similar to the case of probability statements. A probability estimate may be retained even if there are occasionally larger deviations from the relative frequency to be expected. The parallels reach even further. Instead of saying that a sufficiently good approximate confirmation of a hypothesis's control consequences is enough for its acceptance, one declares at times that a metric hypothesis says something only about the limit of the averages just mentioned – just as a probability statement is interpreted as a statement about the limit of relative frequencies, although here too the empirical test is limited in principle to the observation of a more or less approximate agreement between the estimated probability and the relative frequencies in finite series.

If one tries to uphold the classical view about the ultimate refutability of a hypothesis by specifying maximally admissible deviations, one is led into yet other difficulties. As pointed out above, a hypothesis can be tested in very different ways. One can empirically test a hypothesis, for example, by examining just a few of its control sentences or by examining comprehensive systems of its control sentences; the empirical test can extend to control sentences of the same or of a different logical form; and so on. For each of these cases one would have to assign to the hypothesis a special maximum interval of deviations – a construction that obviously conflicts with the procedure practiced in empirical science.

In summary, an empirical sentence is neither ultimately confirmable nor ultimately refutable by a finite set of test results; rather, it can hold its ground only more or less well relative to a certain material. But there are no clear-cut and uniformly applied criteria either for "how well" it minimally has to hold its ground if it is still to be counted as confirmed by the empirical findings in question or for "how badly" it minimally has to hold its ground if it is to be counted as refuted by such findings.

Apart from the degree of agreement between control sentences and control consequences, certain "overarching motives" play an essential role in the decision about the acceptance or rejection of a sentence, such as interest in upholding a universal sentence or regard for securing the greatest simplicity and efficiency of the system S. And for decisions under these motives, there are again no uniform and clear-cut criteria that would permit adequate formal reconstruction.

It should now be sufficiently clear that one can easily transfer the basic idea of the above-mentioned "objection (b)" from finitely interpreted probability statements to all empirical sentences, and also that this objection represents no genuine objection at all, since it rests on an overly schematic view of the character of empirical sentences. This context reveals again that only a finitistic interpretation will really allow us, in accordance with the view of all proponents of an empirical probability theory, to accept probability statements into the class of empirical sentences. For it is only the finitistically, not the transfinitely interpreted, probability statements that the basic idea of "objection (b)" applies to – which, in truth, as we tried to show, gives expression to a characteristic feature of all empirical sentences.

In what follows we will outline a finitistic interpretation of empirical probability statements. This interpretation does not specify the content of probability statements in all details, but only in certain fundamentally important respects. A probability statement (e.g., "$O \xrightarrow{p} P$" in Reichenbach's notation) is an empirical sentence that does not differ in principle from empirical sentences or other kinds; as opposed to giving information – statistical or other – about infinite sets of empirical data, a probability statement determines an indefinitely extensible set of control consequences of various logical forms. An important subset of such control consequences contains consequences of statistical form. These, as empirical sentences, must at any rate be formulated in such a way as to refer to finite observation series – and thus be finitely controllable. The statistical control consequences of a probability statement say that the relative frequency of the event in question falls within a certain interval around the estimated probability value p, where the length of this interval decreases when the length of the statistical series increases. In other words, the statistical control consequences say that for every number n of cases, the following holds:

$$(1) \qquad \left| p - \frac{n_Q}{n} \right| < f(n)$$

Here, p is the probability of the observation result Q (under certain experimental conditions; in Reichenbach's representation, for example, they are taken into account in the specification of O), n_Q is the number of "Q-cases" in n observations under the specified conditions, and $f(x)$ is a monotonically decreasing function of x, a function which asymptomatically approaches the value 0 with increasing x and whose values lie between 0 and 1.

We will not try to specify a particular function of this kind, since the statistical procedure for testing probability statements practiced in empirical science is not so sharply defined that one could "rationally reconstruct" it by means of one and only one function $f(x)$ of this sort.

At any rate, the following will generally hold: $f(x) \geq 1/x$ (for every x), since $|p - n_Q/n| < 1/n$ is equivalent to $|n \cdot p - n_Q| < 1$. But this means that – independent of the total number n – the number of Q-cases equals $n \cdot p$ or, if this is not an integer, that it equals one of the integers "neighboring" $n \cdot p$. Such a far-reaching requirement will in practice hardly ever be taken as a basis for testing probability statements. One can see in connection with this consideration that formulation (1) also avoids the above-mentioned objection (a).

Thus, a probability statement can no more be ultimately confirmed or refuted by a finite set of test results than any other empirical sentence; a probability statement is capable of holding its ground only more or less well, relative to a certain empirical material. To the extent that an empirical test is concerned with the *statistical* control consequences just considered, the decisive factor for an assessment of how well a probability statement holds its ground is primarily the deviation between the value its control consequences specify and the relative frequencies observed. Other factors for such an assessment can be pointed out, but in the scientific use of probability statements, there are no clear-cut criteria that say either under which conditions a probability statement is still to be counted as confirmed in the light of certain empirical findings or under which conditions it is already to be counted as refuted. Once again, overarching motives play an essential role in such a decision.

It is to be noted, however, that control consequences of a statistical form by no means constitute the totality of control sentences of a probability statement. In view of our earlier considerations, several other forms should be recalled. Taking these forms properly into account enables us to loosen up the more narrow, merely statistical interpretation of the probability concept and to carry out the empiricist interpretation of probability statements even more consistently.

To begin with, from a probability statement one can derive such control consequences as have themselves the form of probability statements. These control consequences can then be confronted with control sentences of the form of probability statements that have been otherwise obtained. The theorems of the probability calculus are used in derivations of this sort.

Furthermore, from a probability statement (together with suitable sentence p_x from S), one can derive control consequences of an entirely different form. From statements about the transition probabilities of certain quantum states of a gas, for instance, one can obtain statements about the proportional intensities of certain spectral lines; from the specification of the decay probability of a radioactive substance, one can obtain statements about the number of scintillations a screen shows per minute; and so on.

We can now state more precisely in what respect *the proposal just developed* constitutes *a "finitistic" interpretation of the probability concept*: The finitization relates exclusively to the form of a certain class of control consequences of a probability statement – that is, to the form of statistical control consequences. And, in fact, the central problems of the empirical-statistical probability theory are bound up with control consequences of just this kind.

We did not violate the finitistic maxim when we explicitly said above that *indefinitely many* (finitely-) statistical control consequences were derivable from a probability statement. For what is at issue is not a sentence of empirical science, but a sentence from the syntax of the language of empirical science, a sentence that generally says: From *every* nonanalytic sentence one can derive, in the sense specified above (pp. 113–14), indefinitely many test consequences, even indefinitely many control consequences that are logically independent from each other. When one tests a sentence empirically, one always examines only finitely many of these control consequences. But whereas the limit-interpretation of the probability concept lends a transfinite character to the control questions, we here propose *a finitistic interpretation for the individual control consequences*. It is precisely an interpretation of this sort that is most likely to do justice to the empirical-finite test procedure that science usually applies when testing – like any other hypothesis – a probability statement.

Only the methods of *testing* a probability statement have been at issue in the present section, but we did not ask on the basis of what findings (for instance, of a statistical sort) a probability statement is *advanced*. This question has, however, only apparently been neglected – that is, by its theoretical content, this question is integrated into the problem of testing a hypothesis. For since every hypothesis has an infinite set of logically independent control consequences, no empirical material, no matter how comprehensive it is – however large a set of sentences $g_1, g_2, \ldots, g_m$ – uniquely determines a hypothesis whose control consequences are the g_μ. Rather, the g_μ assign a certain "degree of confirmation" to a tentatively introduced hypothesis h – where this formulation is to be taken as subject to all the above-mentioned qualifications – and one can now look for a hypothesis that holds its ground as well as possible relative to the g_μ. But this "ascent from experience to theory" is by no means uniquely determined, and the genuinely theoretical moments that occur in it belong so obviously to the context of the testing problem that we could dispense with a separate discussion of the above question in the previous sections.

The interpretation of the statistical content of a probability statement proposed above satisfies the conditions developed earlier on. It turns formal probability statements into sentences with empirical content, it has a finite form, and it does in no way refer to the axioms of the probability calculus.

Of course, it does not follow from the last-mentioned factor that the proposed interpretation also turns the "theorems" (formulae) of the formal probability calculus, which surely represent certain relations among probability statements, into sentences with empirical content. Take, for instance, the formula "$p \lor \sim p$" of the propositional calculus; a semantic interpretation of the propositional variable p – say, "It is raining here now" – turns this formula into a sentence *without* empirical content ("It is raining here now, or it is not raining here now"). In the same way, the proposed interpretation could also turn the formulae of the probability calculus into analytic sentences. This possibility is now realized for the theorems of a certain part of the formal probability calculus. One can specify a finite arithmetical model for the above-mentioned axioms of the probability calculus that

Reichenbach has stated. This model can in principle be described as follows. Let n be any fixed natural number. Let $x_1, x_2, \ldots, x_n$ be real numbers, O the set of these numbers, and Q a property that can be meaningfully ascribed to or denied of a real number. The probability statement $O \xrightarrow[p]{} Q$ is now to be interpreted this way: "The number of those numbers x_i in O that have property Q, divided by the total number n, equals p."

This finite-statistical interpretation turns all of Reichenbach's axioms into analytic sentences. It can be immediately inferred from this that the axioms of the probability calculus and also the theorems derivable from them hold analytically for every finite series of empirical observation data which is given as closed, if the probability of an observation result is defined in terms of its relative frequency in the entire finite series. Thus, stated in terms of an example, the addition theorem formulated in Reichenbach's "Axiomatik" (p. 578) as Axiom III, 1 says only this: If in a finite set of casts of a die, 20% of the casts have the result "5" and 14% of the casts have the result "6," then 34% of the set's casts have the result "5 or 6." Similar considerations can be set forth for every other axiom of the system.

If the application of the probability calculus were limited to cases of this kind, then even in a finitistic interpretation the probability calculus would only be a system of analytic sentences and not a theory with empirical content. But in empirical science we make a more far-reaching use of the theorems of the probability calculus. We apply them also to data sequences which are not yet given with all their elements but which are, though finite, still "arbitrarily extendible." The sentences obtained in this way would be analytic if the relative frequencies of all events under consideration remained constant for each extension of the statistical sequence. Admittedly – if only for merely arithmetical reasons – this is not always strictly possible; but at any rate this consideration shows that the approximate validity of formulas under the finite-empirical interpretation is guaranteed at least for those empirical sequences in which the relative frequencies of the results under investigation remain approximately constant. In this respect, the empirical content of that part of the probability calculus that Reichenbach has made precise with the above-mentioned axioms is now established.

Also the "Law of Large Numbers," whose transfinite formulation was shown earlier (p. 99) to be empty, only expresses, in the sense of the finitistic conception proposed, the condition of the approximate constancy of the relative frequencies in statistical series. We can, therefore, say: In every domain of empirical frequency phenomena in which the finitistically interpreted law of large numbers holds, the finitistically interpreted theorems of the probability calculus also hold.

As emphasized earlier (see sec. 3, "Regarding [1]," p. 100), one can deduce from Reichenbach's axioms all those theorems of the classical probability calculus which in von Mises' theory are deducible from Axiom I alone – that is, from the requirement of convergence of the relative frequencies in the sequences under consideration.

But there is a series of theorems – and among those, important theorems of the classical probability calculus – which von Mises proves under the additional assumption of the irregularity requirement. As Reichenbach shows ("Axiomatik," p. 594 ff.), these statements hold in the system of the general formal probability calculus only under certain presuppositions, which are compatible with the axioms of the general theory but do not follow from them. Reichenbach calls the class of those sentences that are derivable from the axioms only under those additional presuppositions (i.e., essentially, the validity of the so-called special multiplication theorem) the "theory of normal sequences." The theory of normal sequences is related to the general formal probability calculus as, say, a theory of equilateral triangles is related to general Euclidean geometry. (Equilateral triangles satisfy certain restrictive presuppositions, those presuppositions which are compatible with the axioms of the geometry but that do not, on the basis of the axioms, hold generally for every triangle.) As Reichenbach shows elsewhere, one can obtain other such specializations of his general theory.

Every special system so obtained is then transformed by the finitistic-empirical interpretation of probability statements into a particular empirical probability theory, which, however, no longer holds for every domain of empirical statistics that satisfies the finitistically interpreted Law of Large Numbers. Which special probabilistic structure a given statistical domain of experience possesses (i.e., which special probability theory holds for this domain) has to be decided by empirical investigations in each case.

Such a decision presupposes again the finitistic interpretation of probability statements. This interpretation, which was developed above in an attempt at a logico-descriptive analysis of the scientific use of probability statements, finally makes possible a consistent development of the leading idea of any empirical-statistical theory. It assigns an empirical content to probability statements and thereby integrates them into the totality of empirical sentences; and, furthermore, it makes possible a treatment of the probability problem (in the sense of von Mises' program) in parallel with the treatment of the geometry problem. The considerations advanced here seem to me to be basically nothing more than a development of certain consequences of the basic methodological principles of any empirical interpretation of the probability concept.

Added in proof (1935) The manuscript of the present article was submitted in October 1934. Therefore, the relevant investigations published since, of which I regard H. Reichenbach's *Wahrscheinlichkeitslehre* (Leyden: Sijthoff, 1935) and K. Popper's *Logik der Forschung* (Vienna: Springer, 1935) as the most important, are not taken into account here. In addition, I would only emphasize that in my view the objections developed above (see pp. 103–4) apply as well to the refined version of Reichenbach's theory of induction advanced in his *Wahrscheinlichkeitslehre*; for in it (especially § § 79 and 80), the limit concept plays essentially the same role as in the theory of induction presented in his "Axiomatik."

Chapter 7

On the Logical Form of Probability Statements*

1. In the following remarks, an attempt will be made to analyze the logical form of the concept of probability and of the statements in which it occurs, and to apply the result of this analysis to some problems concerning the language of empirical science. Our considerations will be based on the modern statistical theory of probability,[1] which is, at present, certainly by far the most satisfactory and consistent basis for a theoretical account of the concept of probability as it is actually used in empirical science as well as in everyday language.

2. According to the statistical theory, each particular probability statement refers to an infinite sequence k of events and to a property P; the probability statement then has the form: "The probability for an element of k to have the property P is p," which, by the statistical interpretation, is equipollent to "The relative

---

* From Carl G. Hempel, "On the Logical Form of Probability-Statements," *Erkenntnis* 7 (1938), 154–60, and "Supplementary Remarks on the Form of Probability-Statements," *Erkenntnis* 7 (1939), 360–3. Reprinted with the kind permission of Kluwer Academic Press.

[1] R. v. Mises, *Wahrfcheinlichkeit, Statiftik und Wahrheit*, 2nd ed. (Vienna: Springer, 1936). H. Reichenbach, 1. *Wahrfcheinlichkeitslehre* (Leiden: Sijthoff, 1935); 2. "Les fondements logiques du calcul des probabilités," *Ann. Inst. H. Poincaré* 7 (1937), 267–348.

For a general survey of the statistical and other types of theories of probability, see E. Nagel, "The Meaning of Probability," *Journal of the American Statistical Association* 31 (1936), 10–26; C. E. Bures, "The Concept of Probability," *Philosophy of Science* 5 (1938), 1–20.

frequency of those among the first n elements of k which have the property P converges toward the limit p as n goes toward infinity." Thus, according to the statistical theory, probability is a three-termed relation between a property (or class), a sequence, and a real number; and each particular probability statement may be given the form "Prob (P, k, p)" ("The probability of P in k is p"). Here, "p" is a real number constant with $0 \leq p \leq 1$; "k" is a functor[2] whose arguments are the natural numbers and whose values are tetrads of real numbers which may be considered as determining the spatio-temporal coordinates of events. However, we need not restrict ourselves here to probability-sequences consisting of events: The values of k may just as well be names of sentences, of numbers, of physical objects, or the like. Finally, "P" is a one-place predicate which must be meaningfully applicable to the elements of k; in other words, the expression "$P(k(n))$" must yield a meaningful sentence if "n" is replaced by any natural number.

The above statistical definition of probability may now be formulated as follows[3]

$$(\mathrm{I}) \qquad Prob(F,h,x) \equiv \lim_{n \to \infty} \frac{nc\{\hat{m}[(m \leq n) \cdot F(h(m))]\}}{n} = x$$

Here, "F," "h," and "x" are *variables* of the types exemplified by the constants "P," "k," and "p"; "m" and "n" are variables for natural numbers; and "nc" is the functor "the cardinal number of."

In addition to the three-place predicate "$Prob$," which may be called the concept of absolute probability, we introduce a four-place predicate "$Probrel$ (F, G, h, x)" – the concept of relative probability of a property G with respect to a property F in a probability-sequence h – by the following definition:

[2] For this and other logical concepts compare R. Carnap, *The Logical Syntax of Language* (London: Kegan Paul, 1937).

[3] Following the way chosen by H. Reichenbach (cf. footnote 1) and E. Kamke (*Einführung in die Wahrfcheinlichkeitstheorie* [Leipzig: Hirzel, 1932]), we leave aside, in the general definition of probability, any supplementary condition concerning the so-called irregularity of the distribution of F and $\sim F$ in h; the inclusion of such restrictions into the definition would, however, not affect the result that "$Prob$" is a three-place relation of the type indicated above.

$$Probrel(F,G,h,x) \equiv$$

$$(\text{II}) \qquad \lim_{n\to\infty} \frac{nc\{\hat{m}\,[(m \leq n)\cdot F(h(m))\cdot G(h(m))]\}}{nc\{\hat{m}\,[(m \leq n)\cdot F(h(m))]\}} = x$$

The concepts of absolute and of relative probability might also have been introduced as a two-place functor "prob" and a three-place functor "probrel" respectively, their values being real numbers. To the statement "*Prob* (*P*, *k*, *p*)," for example, there would correspond the statement "*prob* (*P*, *k*) = *p*." However, this representation requires special precautions, since the defining limit and, therefore, the probability does not exist for *all* possible values of the argument.

3. In his theory of probability, H. Reichenbach represents particular probability statements concerning *relative* probabilities (let us call them *particular relative probability statements*) in the form

$$(\text{III}) \qquad (i)\,(x_i \in O \xrightarrow[p]{} y_i \in P)$$

Here, O and P are properties (classes), p is a real number, $\longrightarrow$ is the so-called probability implication, and x_i and y_i run through sequences correlated by the index i, such as the throws of a die and the results obtained by the respective throws.

Now, the representation of a particular relative probability statement which has been developed in (II) seems to involve certain advantages as compared with the schema (III):

a. The symbols "x_i," "y_i," which Reichenbach calls variables, are not really individual variables, but constant functors. Indeed, substituting individual constants – say "a" and "b" – for them would transform (III) into an expression which is meaningless in Reichenbach's language. That "x_i," "y_i" cannot be two variables may already be guessed from the fact that there is only *one* binding operator "(i)" for both of them.

b. Furthermore, in cases of the above kind, we can always do with *one* sequence (hence with one functor) instead of with two of them. (Instead of saying, "The *result* of the tenth throw has the property (P) of yielding a five," we may just as well say, "The *tenth throw* has the property (Q) that its result is a five.")

Together, (a) and (b) suggest that one choose, instead of (III), a formula like "(i) $\{k \, (i) \in O \xrightarrow{p} k \, (i) \in Q\}$." Now, here, making use of an integer-variable "i" is not adequate, since a substitution of a constant value – say, "3" – would again lead to a meaningless expression. Thus, what (III) asserts is a relation between the sequence (now designated by "k"), the classes O and Q, and the real number p; this leads us to the representation of the statement in terms of a four-termed predicate which may be written "$\xrightarrow{\;\;}$" or, as we did above, "*Probrel.*" So we come to the expression "*Probrel* (O, Q, k, p)" instead of (III).

As equivalent expressions for (III), Reichenbach introduces the formulae "$O \xrightarrow{p} P$" and "$W \, (O, P) = p$." Since, as we just saw, "$\xrightarrow{\;\;}$" has to be construed as a four-place predicate, this briefer symbolism is incomplete. It does not contain any reference to the sequence under consideration. (It may, however, be used, if we assume that all the formulae written in a certain context, in terms of the brief symbolism, refer to one and the same sequence.) An important consequence of this point will be discussed in (5).

4. As we have already mentioned, the members of a sequence to which a probability statement refers may consist of *sentences* belonging to a given language L, such as that of empirical science; we may then examine the probability for a sentence of such a sequence l to be true (in L).[4] If this probability exists and is equal to p, we may establish the statement "*Prob* $(True_L, l, p)$," and we may speak of a *semantical application of* the concept of *probability*. A particularly important case of this kind has been investigated by Reichenbach.[5] We can now describe this case in the following way. To every particular (absolute) probability statement in L –

---

[4] An exact theory of the concept of truth with respect to a given (formalized) language has been established by A. Tarski; compare his paper, "Der Wahrheitsbegriff in den formalisierten Sprachen," *Studia Philosophica* (1936), 261–405; for a more general account of this theory and its significance as a foundation for a rigorous theory of semantics, compare M. Kokoszynska, "Über den absoluten Wahrheitsbegriff und einige andere semantische Begriffe," *Erkenntnis* 6 (1936), 143–65.

[5] Compare (in addition to his book mentioned in footnote 1): (1) "On Probability and Induction," *Philosophy of Science* 5 (1938), 21–45; (2) *Experience and Prediction* (Chicago: The University of Chicago Press, 1938), §33.

say, "*Prob* (*P, k, p*)" – there is an equipollent probability statement in the semantics of *L* which has the form "*Prob* (*True$_L$, l, p*)" (here, *l* is the following sequence of sentences: "*P* (*k* (1))," "*P* (*k* (2))," "*P* (*k* (3))," . . .). In fact, the probability of *P* in *k* is *p* if and only if the probability of truth in *l* is *p*. Reichenbach even speaks, in this context, of an *isomorphism* of the concept "*Prob*" as applied in *L*, and "*Prob*" as applied in the semantics of *L*. But an isomorphism does not really exist, since the domains of the two concepts are not even correlated by a one-to-one relation. In fact, to the same sequence *k* different sequences $l_1, l_2, \ldots$ will correspond, if different properties $P_1, P_2, \ldots$ are considered; in other words, *l* is determined by both *k* and *P*, not by *k* alone. Furthermore, as one easily verifies, there is no isomorphism at all with respect to *relative* probability statements; it is not possible to coordinate to a relative probability statement in *L* – say, "*Probrel* (*P, Q, k, p*)" – an equipollent relative probability statement in semantical terms, one that is analogous to that considered above.

5. Finally, the analysis sketched in sections 2 and 3 has a consequence which seems to be important for the investigation of the logical structure of empirical science. As we saw, probability as dealt with in the statistical theory is a relation, one of whose terms is an infinite sequence; unless the sequence in question is explicitly indicated, the probability statement is essentially incomplete, and the mathematical theory of probability in its statistical form cannot be applied to it. On the other hand, in the probability statements occurring in empirical science, a reference to a special sequence is practically never made. This shows that probability statements of the forms "*Prob* (*P, k, p*)" or (III) or "*Probrel* (*O, P, k, p*)" do not yet render adequately the logical form of the probability statements established in empirical science. Let us consider an example. If *d* is a certain die, then, in the language of empirical science, the following statement S_1 may be established: "The probability of throwing a five with *d* (by means of such and such a tossing mechanism – e.g., the hand of a gambler) is 0.16." In the statistical theory, however, such a statement would be incomplete; it would require the indication of a sequence of events with respect to which that probability is asserted to be 0.16, and only to complete probability statements indicating the

sequence to which they refer does the statistical theory of probability apply. At first look, it seems as if the meaning of the empirical statement could be expressed in terms of the statistical theory by this sentence (S_2): "The probability of the result five is 0.16 in *every* sequence h which fulfills the condition that each of its elements is a throw with d." However, S_2, far from being equipollent to S_1, can be shown to be contradictory. If namely k is a certain sequence which fulfills the above condition and confers to the result five the probability 0.16, then obviously k contains a proper subsequence k_1 in which all members have the property of yielding the result five – and which, therefore, confers to that result the probability 1, hence not 0.16; and nevertheless k_1, like k, fulfills the above condition. Analogously, it can be seen that for any desired probability-value x, k contains a subset which gives to the result five just the probability x.

Quite an analogous consideration applies to the probability statements which are established in the kinetic theory of gases, in the theory of radioactivity, or in biological and social statistics. These statements also do not contain a characterization as to the sequences to which the statement refers; and the question is just how to reformulate the statements in terms of the statistical theory. This point is particularly important for any attempt at interpreting – as Reichenbach proposes[6] – every empirical law as a probability statement. For this interpretation amounts, in the simplest case, to replacing a universal implication – say, "$(x) (P (x) \supset Q (x))$" – by a relative probability statement – say, "*Probrel* (P, Q, k, p)" – and this procedure requires, apart from a probability value p, the explicit indication of a sequence k, a fact which seems not to have been noticed so far.

On account of the preceding considerations, it seems to be necessary to include, in a formalization of an empirical probability statement, a characterization of a class of sequences to which the statement is meant to refer. The representation of an absolute empirical probability statement would then assume the form

---

[6] Compare "Über Induktion und Wahrscheinlichkeit," *Erkenntnis* 5 (1935), 267–84.

(IVa) $\quad$ "$(h)(C(h) \supset Prob(P,h,p))$"

and that of a relative probability statement would be

(IVb) $\quad$ "$(h)(C(h) \supset Probrel(P,Q,h,p))$"

Here, C is the condition or group of conditions characterizing those sequences in which P (or P with respect to Q) has the probability p. The content of the conditions "C" will depend, in each case, upon the property P (or, upon P and Q, respectively), so it cannot be stated in a general way for all empirical probability statements. But even for a given concrete case, it seems to be rather difficult to formulate explicitly the conditions "C" in question. Thus, for example, one might try to reformulate the sentence "The probability of throwing a five with the die d is 0.16" by the following statement which has the form of (IVa):

> The probability of obtaining the result five is 0.16 in every sequence h which fulfills the following conditions (C): that the first element of h is some throw with d performed by such and such a tossing mechanism, and that for any n, the nth element of h is a throw with d, performed $n - I$ seconds after the first throw.

However, it is obvious that the class of sequences thus determined is too narrow; the assertion "The probability of getting a five with d is 0.16" is certainly meant to apply to many more sequences obtainable by means of d – although not to all of them. In fact, an adequate definition of C would have to include all but those sequences which are defined by direct reference to their results (the inclusion of the latter would lead us into contradictions, as we have seen before). Thus, the problem of defining C turns out to be very closely related to that which v. Mises undertook to solve by means of his principle of the excluded gambling system – namely the problem of characterizing statistical irregularity.[7] This problem arises here again – in a new context – for each of the various systematizations of the statistical theory which have so far been developed (even for v. Mises' theory, which contains the principle of the excluded gambling system as

---

[7] Compare footnote 3.

an axiom), because all of them use the concept of probability as a predicate which has the structure pointed out in (2), and therefore contains one more argument than the probability statements used in empirical science.

SUPPLEMENTARY REMARKS

6. Some of the comments made in the discussion referred to a well-known difficulty which is connected with the statistical theory of probability – but which was not intended to be dealt with in my paper. That difficulty arises from defining the probability of a characteristic *P* as the *limit* of its relative frequency. In fact, it has repeatedly been noticed that because of this "transfinite" definition, we can never find the probability of *P*; neither can we verify empirically an assertion about its value, since we cannot survey infinite sequences in our finite experience. Putting it somewhat more precisely, the difficulty is that a probability statement about *P*, according to that transfinite definition of probability, is not susceptible of an empirical verification or falsification, for any empirical evidence concerning the frequency of the occurrence of *P* would necessarily be restricted to *finite* sets of events; and whatever the observed frequency of *P* in a finite set may be, it is compatible with whatever limiting value one wants to assume. Even if, for example, 1,000 consecutive throws of a die yield always the result five, the relative frequency of that characteristic may yet, without any inconsistency, be assumed to be $\frac{1}{6}$.

Thus, as amenability to an empirical test is a constitutive characteristic for an empirical sentence, the transfinite definition of probability seems to deprive probability statements of any empirical content.[8] As a consequence, several authors have claimed – and this view was shared by some of the speakers in the discussion – that the mentioned definition must be abandoned, for, the argument runs, it is impossible to apply a theory

---

[8] For a more detailed discussion of this problem, compare C. G. Hempel, "Über den Gehalt von Wahrscheinlichkeitsaussagen," *Erkenntnis* 5 (1935), 228–60 (Chapter 1 here); and the publications mentioned in footnote 1 earlier.

which makes essential use of such concepts as "infinite sequence" and "limiting value."

However, there are certain reasons which do not seem to speak in favor of such a general verdict:

a. Many concepts in physical theory are of the incriminated kind; this holds, for example, for such concepts as "speed" and "acceleration," which may be introduced as certain limiting values. (A closer analysis reveals that they are transfinite in an even more serious way: Their definitions refer to infinite classes of denumerable sequences.) Nevertheless, the theories which employ these concepts yield most fruitful applications. This is due to the fact that there are certain universal laws which connect the concepts in question with others, such as "impulse," "force," etc., in such a way that, for example, the acceleration of a body may be determined by measuring a certain force – which may be done by means of a balance, without any reference to infinite sequences of any kind.[9] This circumstance makes statements about the magnitude of a speed or of an acceleration accessible to empirical tests. Analogous devices are sometimes applicable to probability statements; thus, for example, an assertion about the probability for an electron of a sodium-atom to jump from a certain orbit to another may be tested by measuring the intensity of a spectral line.

b. Apart from this point, there are also certain general considerations which, as Carnap[10] has recently pointed out, suggest a way to determine the domain of the empirically meaningful sentences in such a liberal way as to comprehend the probability statements of the transfinite type mentioned before. In fact, a more detailed analysis of the concept "limiting-value" shows that those probability statements, though not immediately verifiable or refutable by empirical observations, are nevertheless con-

---

[9] The fact that the exact position of the pointer will in general be determined by reference to a *series* of observations need not be taken into consideration here, because it does not eliminite the difference between the case in question and that of determining a probability.

[10] See R. Carnap, "Testability and Meaning," *Philosophy of Science* 3 (1936), 419–71, and 4 (1937), 1–40, particularly the last section of that paper.

nected – in a more indirect way, it is true – with sentences expressing empirical observations. Thus, they are accessible to empirical testing.

At any event, it has to be considered as a matter of convention, not as a question of fact, whether limiting-concepts and sentences of the form of tranfinite probability statements are to be admitted in empirical science – or to be excluded from it.

7. The considerations developed in sections 1–5 above intentionally leave aside the discussion of this problem. Their purpose is to describe the logical structure which probability statements possess according to the transfinite statistical interpretation as well as to point out a difficulty which results from this interpretation. That difficulty is independent of the one mentioned before, and it seems not to have been noticed so far. Roughly, it may be expressed as follows. Let us suppose that there were no difficulties in empirically determining the limit of the relative frequency of P in an infinite sequence. Then we should be in a position to make assertions of the following kind: "In the sequence k, the limit of the relative frequency of the result five is 0.16," or "k is a 0.16-probability-sequence for the result five"; and the statistical theory would tell us how to derive the probabilities of other results in k from given probabilities, etc. However, all this would not enable us to give a satisfactory account of the meaning of probability statements as used in empirical science, for these statements do not speak of sequences at all. Thus, for example the statement S ("The probability of throwing a five with a homogeneous and symmetrical die is $\frac{1}{6}$") does not contain any indication as to what particular sequence is asserted to be a $\frac{1}{6}$-probability-sequence with respect to the result five. How, then, can we express its meaning in terms of the statistical theory?

Even if we assume v. Mises' rigorous form of the statistical theory and suppose that we could decide by empirical observation whether a sequence is a collective in v. Mises' sense – even then the question remains the same: Which sequences does the statement S assert to be collectives giving the probability $\frac{1}{6}$ to the result five? Not just every sequence consisting of throws of a normal die can be asserted to be such a collective, since, as has been shown in the first paragraph of sec. 5 above, there are certain sequences of that kind which yield exclusively the result five, as well as others which never yield it at all; and these sequences are certainly no collectives.

8. The difficulty under consideration is not restricted to those forms of the statistical theory which are based on the above transfinite definition of probability; it arises for "finitistic" statistical theories of probability[11] as well. In fact, the latter define the probability of a characteristic P some way or other by means of its frequency in *finite* classes (instead of in infinite sequences or collectives); empirical probability statements, however, such as that about throwing a five with a normal die, do not refer to particular finite classes of events (no more than to particular collectives). Thus, the considerations outlined in the preceding section apply, *mutatis mutandis*, also to finitistic statistical interpretations of probability.

---

[11] A finitistic interpretation of the concept "probability" as used in empirical science has been proposed in the paper mentioned in footnote 8 of the present remarks. This interpretation does not depend upon whether the mathematical calculus of probability is given a transfinite or a finite form. A representation of the *calculus* of probability in a *finitistic* form – namely, as a theory of finite collectives – has been developed by Johannes Blume. Compare his papers: (1) "Mathematische Begründung und Entwicklung einer Wahrscheinlichkeitsrechnung mit finiten Kollektiven," *Zeitschrift für Physik* 92 (1934), 232–52; (2) "Zur Anwendung der Wahrscheinlichkeitsrechnung finiter Kollektive," *Zeitschrift für physik* 94 (1935), 192–203.

Chapter 8

A Definition of "Degree of Confirmation"*

[i.e. what does this MEAN, concretely?]

1. The Problem. The concept of confirmation of a hypothesis by empirical evidence is of fundamental importance in the methodology of empirical science. For, first of all, a sentence cannot even be considered as expressing an empirical hypothesis at all unless it is theoretically capable of confirmation or disconfirmation, that is, unless the kind of evidence can be characterized whose occurrence would confirm, or disconfirm, the sentence in question. And second, the acceptance or rejection of a sentence which does represent an empirical hypothesis is determined, in scientific procedure, by the degree to which it is confirmed by relevant evidence.

The preceding remarks, however, are meant only as accounts of methodological tendencies and are not intended to imply the existence of clear-cut criteria by means of which the scientist can decide whether – or, in quantitative terms, to what degree – a given hypothesis is confirmed by certain data. For indeed, no general and objective criteria of this kind are at present available; in other words, no general definition of the concept of confirmation has been developed so far. This is a remarkable fact in view of the importance of the concept concerned, and the question naturally suggests itself whether it is at all possible to set up

* From Carl Hempel (with Paul Oppenheim), "A Definition of 'Degree of Confirmation,'" *Philosophy of Science* 12 (1945), 98–115. Reprinted by permission of the University of Chicago Press.

[QUOTE! PREF]

But:

adequate general criteria of confirmation, or whether it may not rather be necessary to leave the decision in matters of confirmation to the intuitive appraisal of the scientist.

This latter alternative would be highly unsatisfactory; for first, it would clearly jeopardize the objectivity – in the sense of intersubjectivity – of scientific procedure. Second, it would run counter to a view of confirmation which is now widely accepted. According to this view, statements about confirmation assert nothing regarding an observer's subjective appraisal of the soundness of a hypothesis; rather, they concern a certain objective relation between a hypothesis and the empirical evidence with which it is confronted. This relation depends exclusively on the content of the hypothesis and of the evidence, and it is of a purely logical character in the sense that once a hypothesis and a description of certain observational findings are given, no further empirical investigation is needed to determine whether, or to what degree, the evidence confirms the hypothesis; the decision is a matter exclusively of certain logical criteria which form the subject matter of a formal discipline which might be called inductive logic.

Of course, the widespread acceptance of this view does not prove that it is sound and that the program implicit in it can actually be carried out. The best – and perhaps the only – method of settling the issue seems to consist in actually constructing an explicit and general definition and theory of confirmation. To do this is the purpose of this article. It is intended to present in outline, and with emphasis on the general methodological issues, a theory of confirmation which was developed by the present authors jointly with Dr. Olaf Helmer.[1]

---

[1] A detailed technical exposition of the theory is given by Olaf Helmer and Paul Oppenheim in "A Syntactical Definition of Probability and Degree of Confirmation," *The Journal of Symbolic Logic* 10 (1945), 25–60.

The present issue of *Philosophy of Science* contains an article ("On Inductive Logic," *Philosophy of Science* 12 [1945], pp. 72–97) by Professor Rudolf Carnap which likewise sets forth a definition and theory of confirmation. The approach to the problem which is to be developed in the present paper is independent of Professor Carnap's and differs from it in various respects. Some of the points of difference will be exhibited subsequently as the occasion arises. We wish to

A Definition of "Degree of Confirmation"

As is illustrated by the terminology used in the preceding discussion, the concept of confirmation may be construed as a metrical (quantitative) as well as a purely classificatory (qualitative) concept. These two different forms are exemplified, respectively, in the phrases "The degree of confirmation of the hypothesis H relatively to the evidence E is such and such" and "The evidence E is confirming (disconfirming, irrelevant) for the hypothesis H." The theory here to be presented deals with the metrical concept of confirmation; its objective is to construct a definition of the concept of degree of confirmation and to derive, from this definition, a number of consequences, which may be called theorems of inductive logic.[2]

2. *The Language* L. The planned definition or "rational reconstruction" of the concept of degree of confirmation in precise terms can be carried out only on the assumption that all the hypotheses to be considered and all the reports on observational data are formulated as sentences of a language, L, whose logical structure and means of expression are precisely determined. More specifically, we shall presuppose that L contains the following means of expression, which will be symbolized, but for minor and obvious changes, in accordance with the familiar notation of *Principia Mathematica*:

2.1 The statement connectives "~" ("it is not the case that"), "·" ("and"), "∨" ("or"), "⊃" ("if . . . then")

express our thanks to Professor Carnap for valuable comments he made in the course of an exchange of ideas on the two different studies of confirmation.
We also wish to thank Dr. Kurt Gödel for his stimulating remarks.

[2] For a definition and theory of the classificatory concept of confirmation, see the following two articles by Carl G. Hempel: "A Purely Syntactical Definition of Confirmation," *The Journal of Symbolic Logic* 8 (1943), 122–43; "Studies in the Logic of Confirmation," *Mind* 54 (1945), 1–26, and 97–121.

The technical term "confirmation" should not be construed in the sense of "verification" – an interpretation which would preclude, for example, its application to a hypothesis about an event which is temporally posterior to the data included in the evidence. Rather, as is suggested by the root "firm," the confirmation of a hypothesis should be understood as a strengthening of the confidence that can rationally be placed in the hypothesis.

2.2   Parentheses

2.3   Individual constants, that is, names of individual objects (which may be physical bodies, events, space-time regions, or the like), "a_1", "a_2", "a_3", . . . (Here, quotation marks are used to form names of the material between them so that commas separating such names are left outside the quotes. –Editor) The number of individual constants in L may be finite or denumerably infinite.

2.4   Some finite number p of one-place predicates, that is, names of properties which any one of the individuals referred to under 2.3 may or may not have:

$$"P_1", "P_2", "P_3", \ldots, "P_p"$$

These predicates are undefined concepts in L; we shall, therefore, refer to them as the primitive predicates of L and to their designata as the primitive properties referred to in L.

2.5   Individual variables: "x", "y", "z", . . . in any number.

2.6   The symbols of universal and existential quantification, as illustrated in "$(x)P_1x$" and "$(Ey)P_1y$".

We further assume that these symbols can be combined in the customary ways to form sentences in L[3] and that the usual rules of deductive inference govern the language L.

Briefly, then, we assume that L has the logical structure of the so-called lower functional calculus without identity sign and that is it restricted to property terms only. These assumptions involve a considerable oversimplification from the viewpoint of the practical applicability of the theory here to be presented, for the language of empirical science includes a great deal of additional logical apparatus, such as relation terms, expressions denoting quantitative magnitudes, etc. However, in appraising the significance of this restriction, the following points might well be borne in mind: 1. In the case of the concept of degree of

---

[3] Illustrations: "$(P_1a_1 \lor P_2a_1) \supset (P_3a_1 \cdot {\sim}P_4a_1)$" stands for "If a_1 has at least one of the properties P_1, P_2, then it has the property P_3 but not the property P_4"; "${\sim}(x)P_1x \supset (Ex){\sim}P_1x$" stands for "If it is not the case that all objects have the property P_1, then there is at least one object which does not have the property P_1."

confirmation, for which no explicit definition has been available at all and for which even the theoretical possibility of a definition has been subject to serious doubt, it seems to be a significant achievement if such a definition can be provided, even if its applicability is restricted to languages of a comparatively simple structure. 2. While the means of expression of L are relatively limited, they still go beyond the logical machinery which forms the subject matter of traditional Aristotelian logic. 3. The formulation of a definition for languages of our restricted type may serve as a guide in the construction of an extension of the definition to more complex language forms.

3. *Some Auxiliary Concepts.* By an *atomic sentence* we shall understand any sentence of the kind illustrated by "P_1a," which ascribes a primitive property to some individual. Any sentence such as "$(P_1a \supset \sim P_2a). P_1b$," which contains no quantifiers, will be called a *molecular sentence*.

Let "Ma_i" be short for some molecular sentence such as "$P_1a_i \cdot (P_2a_i \vee \sim P_3a_i)$," which contains only one individual constant, "a_i"; then we shall say that "M" designates a *molecular property* – in the example, the property $P_1 \cdot (P_2 \vee \sim P_3)$. By a *statistic* we shall understand any sentence of the type

$$\pm Ma_i \cdot \pm Ma_j \cdot \ldots \cdot \pm Ma_t$$

where the constants "a_i", "a_j", . . . , "a_t" are all different from one another and where the symbol "$\pm$" indicates that any one of the components may be either negated or unnegated. If we wish to indicate specifically the molecular property about whose incidence the statistic reports, we shall call a sentence of the above kind an *M-statistic*.

A sentence which contains at least one quantifier will be called a *general sentence*; in particular, all general laws, such as "$(x)(P_1x \supset P_2x)$," are general sentences.

By means of the p primitive predicates, we can form exactly $k = 2^p$ different conjunctions of the following kind: Each conjunction consists of exactly p terms; the first term is either "P_1" or "$\sim P_1$", the second is either "P_2" or "$\sim P_2$," and so on; finally, the pth term is either "P_p" or "$\sim P_p$". We call these expressions Q-

expressions and, in lexicographic order, abbreviate them by "Q_1", "Q_2", ..., "Q_k". These Q-expressions designate certain molecular properties, which we shall call *Q-properties*. Alternatively, we may also say that each Q-expression designates a class, namely the class of all those individuals which have the Q-property in question. The classes designated by the Q-expressions clearly are mutually exclusive and exhaustive: Every object belongs to one and only one of them. Moreover, they are the narrowest classes which can be characterized in L at all (except for the null class, which is designated, for example, by "$P_1 \cdot \sim P_1$"); for brevity, we shall refer to them as *(L-)cells*. In intuitive terms, we may say that if for a given individual we know to which L-cell it belongs, then we know everything about that individual that can be said in L at all; it is completely determined – relatively to the means of expression of L.

4. A Model Language and a Model World. Our assumptions and definitions for L concern only the logical structure of that language and leave room for considerable variation in material content. For illustrative purposes, it will be useful to be able to refer to a specific model L_W of such a language and to a "model world" W of which it speaks.

Let us assume that the individuals $a_1, a_2, a_3, \ldots$ of which L_W speaks are physical objects, and that L_W contains just two primitive predicates, "Blue" and "Round". Then L_W determines exactly four cells, $Q_1 =$ Blue·Round, $Q_2 =$ Blue·~Round, $Q_3 =$ ~Blue·Round, $Q_4 =$ ~Blue·~Round. All the hypotheses and evidence sentences expressible in L refer exclusively to the characteristics of blueness and roundness of the objects in W. Thus, for example, the evidence sentence E might report, in the form of a statistic, on a sample of individuals in the following manner:

4.1 $E =$ "Blue a_1·Round a_1·Blue a_2·Round a_2·
~(Blue a_3·Round a_3)·Blue a_4·Round a_4"

and the hypothesis might be

4.2 $H =$ "Blue a_5·Round a_5"

In this case, E reports on four objects, three of which were found to be blue and round, while one was not; and *H* asserts that a fifth object, not yet examined (i.e., not referred to in *E*), will be blue and round. The question then arises: What degree of confirmation shall be assigned to *H* on the basis of *E*? We shall return to this case in the following section.

5. *Restatement of the Problem.* Our basic problem can now be restated as follows: to define, in purely logical terms, the concept "degree of confirmation of *H* relatively to *E*" – or briefly, "*dc*(*H*, *E*)" – where *H* and *E* are sentences in a language *L* of the structure characterized in section 2 and where *E* is not contradictory.

The restriction of *E* to logically consistent sentences is justifiable on pragmatic grounds: No scientist would consider a contradictory "evidence sentence" as a possible basis for the appraisal of the soundness of an empirical hypothesis. But the same restriction is demanded also, and more urgently, by considerations of generality and simplicity concerning the formal theory of confirmation which is to be based on our definition. We shall try to define *dc* in such a way that the following conditions, among others, are generally satisfied:[4]

5.1 $dc(H, E) + dc(\sim H, E) = 1$
5.2 If *H* is a logical consequence of *E*, then $dc(H, E) = 1$

But these requirements cannot be generally satisfied unless *E* is noncontradictory, for if *E* is a contradictory sentence, then any hypothesis *H* and its denial ~H are consequences of *E*, and, therefore, by virtue of 5.2, both have the *dc* 1 with respect to *E*. Hence $dc(H, E) + dc(\sim H, E) = 2$, which contradicts 5.1.[5]

As the illustration 4.1 suggests, it might seem natural further to restrict *E* by the requirement that it has to be a molecular sentence, for in practice, *E* will usually consist in a report on a finite number of observational findings. However, it also happens in science that the evidence adduced in support of a hypothesis

---

[4] Here and at some later places we use statement connective symbols autonymously (i.e., roughly speaking, as designations of the same symbols in the "object language" *L*).

[5] This argument was suggested by Professor Carnap.

(such as Newton's law of gravitation) consists of general laws (such as Kepler's and Galileo's laws); and in the interest of the greatest possible adequacy and comprehensiveness of our definition, we shall, therefore, allow E to be any noncontradictory sentence in L. The sentence H, which represents the hypothesis under consideration, will be subject to no restrictions whatsoever; even analytic and contradictory hypotheses will be permitted; in these latter two cases, no matter what the evidence may be, the dc will yield the values 1 or 0, respectively, provided that dc is defined in such a way as to satisfy 5.1 and 5.2.

One of the guiding ideas in our attempt to construct a definition of confirmation will be to evaluate the soundness of a prediction in terms of the relative frequency of similar occurrences in the past. This principle appears to be definitely in accordance with scientific procedure, and it provides certain clues for a general definition of dc. Thus, for example, in the case stated in 4.1 and 4.2, we shall want $dc(H, E)$ to be equal to $\frac{3}{4}$. And more generally, we shall want our definition to satisfy the following condition:

5.3   If E is an M-statistic and H a sentence ascribing the property M to an object not mentioned in E, then $dc(H, E)$ is to be the relative frequency of the occurrence of "M" in E.[6]

This rule is closely related to Reichenbach's rule of induction. This is no coincidence, for Reichenbach's theory, too, aims at giving a strictly empiricist account of the inductive procedure of science.[7] The applicability of the rule 5.3 is obviously restricted to the case where E is a statistic and H has the special form just described. And since we cannot presuppose that in science H and E are generally of this very special type, it becomes an important problem to find a rule whose scope will include also more complex forms of H and of E. In fact, this rule will have to be applicable to any H and any consistent E in L; and in cases of the

---

[6] On this point, see also sections 10 and 16 in Professor Carnap's article (see footnote 1).

[7] Compare Hans Reichenbach, *Wahrscheinlichkeitslehre* (Leiden: Sijthoff, 1935), especially §§75–80, and *Experience and Prediction* (Chicago: University of Chicago Press, 1935), Chapter V.

special type just considered, it will have to yield that value of dc which is stipulated in 5.3. We shall now develop, in a number of steps, the ideas which lead to a definition of the desired kind.

6. *Frequency Distributions.* We have seen that for a given language L, the p primitive predicates determine $k = 2^p$ cells $Q_1, Q_2, \ldots, Q_k$. Each one of these cells may be occupied or empty (i.e., there may or there may not be individuals having the property which characterizes the elements of that cell). Whether a given cell is empty – and if not, how many objects it contains – is of course an empirical question and not a matter to be settled by logic. At any rate, if the number of all objects is finite – say N – then each cell Q_s has a certain occupancy number (i.e., number of elements), N_s, and a certain relative frequency $q_s = N_s/N$. Obviously, $N_1 + N_2 + \ldots + N_k = N$, and

6.1 $$q_1 + q_2 + \ldots + q_k = 1$$

If the class of all individuals is infinite – and here we restrict ourselves to the case of a denumerably infinite set of objects – then we shall assume that they are arranged in a fixed sequence, and by q_s we shall now generally understand the limit of the relative frequency with which elements belonging to cell Q_s occur in that sequence.

Now, while we do not actually know the values q_s, we may nevertheless consider certain hypothetically assumed values for them and develop the consequences of such an assumption. By a frequency distribution Δ in L, we shall understand any assignment of nonnegative numbers $q_1, q_2, \ldots q_k$ to the cells $Q_1, Q_2, \ldots Q_k$ in such a fashion that 6.1 is satisfied. We shall briefly characterize such a distribution by the following kind of notation:

$$\Delta = \{q_1, q_2, \cdots q_k\}$$

It follows immediately that in every Δ,

6.2 $$0 \leqq q_s \leqq 1 \quad (s = 1, 2, \cdots, k)$$

Example: In the case of L_W, one of the infinitely many possible frequency distributions is $\Delta = \{\frac{1}{2}, \frac{1}{3}, 0, \frac{1}{6}\}$, which represents the case where one half of all objects are blue and round, one third of them blue and not round, none of them round and not blue, and one-sixth of them neither blue nor round.[8]

7. *Probability of a Hypothesis.* If a fixed frequency distribution Δ is given or hypothetically assumed for the cells determined by L, then it is possible to define a concept "$pr(H, E, \Delta)$" – in words: "the probability of H relative to E according to the distribution Δ" – which we shall then use to define dc (H, E).

The meaning of this probability concept will first be explained by reference to our model. Let us consider the process of establishing evidence sentences – and of testing hypotheses by means of them – in analogy to that of drawing samples from an urn and using the evidence thus obtained for the test of certain hypotheses. The latter may concern either the distribution of certain characteristics over the whole population of the urn or the occurrence or nonoccurrence of certain characteristics in objects subsequently to be drawn from the urn. For the sake of simplicity we shall assume from now on that the totality of all objects to which L refers is denumerably infinite. [This does not necessarily mean that L contains infinitely many individual constants, but it does mean that the universal and existential quantifiers occurring in the general sentences of L refer to an infinite domain.] Now let us imagine that for our model world W we are given the frequencies associated with the four cells determined by the two predicates of L_W; let this distribution be $\Delta_1 = \{\frac{1}{2}, \frac{1}{3}, 0, \frac{1}{6}\}$. Suppose further that the hypothesis $H_1 = $ "Blue $a_1 \cdot$ Round a_1" is under consideration. We wish to show that a definite probability $pr(H_1, E, \Delta_1)$ can be assigned to H_1 with respect to any given E (with a restriction to be mentioned subsequently), according to the frequency distribution Δ_1.

---

[8] Note that distributions cannot be characterized in L and that, therefore, they cannot form the content of any hypothesis that may be formulated in L. We speak about them in a suitable metalanguage for L. In our case, this metalanguage is English, supplemented by a number of symbols, such as "H", "E", "q_1", "q_2", ... , "Δ", etc. It might be well to emphasize at this point that the definition and the entire theory of dc for L is formulated in that metalanguage, not in L itself. In the metalanguage, we speak about the sentences of L and about the degrees to which certain sentences confirm others.

I. We first consider the case where no information besides Δ_1 is available. In this case, E may be taken to be some analytic sentence, say "Blue a_1 ∨ ~Blue a_1", which we shall designate by "T". Thus, we are concerned with an explanation of pr (H_1, T, Δ_1) – that is, the probability of H_1 according to the frequency distribution Δ_1. We shall construe the problem of defining this magnitude in strict analogy to the following question: Given the distribution Δ_1 for the population of an urn W, what is the probability that the first object drawn will be both blue and round? Since there is no discrimination among the objects except in terms of their properties referred to in L_W, this latter probability will be the same as the probability that some object, chosen at random from the urn, will be both blue and round. The latter probability, however, is uniquely determined by the given distribution: It is the relative frequency assigned to Q_1 in Δ_1. In our case, therefore, $pr(H_1, T, \Delta_1) = \frac{1}{2}$. Now let $H_2 =$ "Blue a_1," which is logically equivalent to "(Blue a_1·Round a_1)∨(Blue a_1· ~Round a_1)." This sentence asserts that a_1 belongs to one of two cells whose occupancy frequencies, according to Δ_1, are $\frac{1}{2}$ and $\frac{1}{3}$, respectively. Since the cells are mutually exclusive, we set pr (H_2, T, Δ_1) = $\frac{1}{2}$ + $\frac{1}{3}$. Similarly, for $H_3 =$ "~Blue a_2", $pr(H_3, T, \Delta_1) = \frac{1}{6}$. Finally, let $H_4 =$ "Blue a_1·Blue a_2·~Blue a_3"; then we set $pr(H_4, T, \Delta_1) = \frac{5}{6} \cdot \frac{5}{6} \cdot \frac{1}{6}$.[9]

After these illustrations, we shall now outline a general method of determining $pr(H, T, \Delta)$ for any given H and Δ. For this purpose, we introduce an auxiliary concept. By a *perfect description*, we shall understand a conjunction each of whose terms assigns some particular individual to some definite L-cell, and in which no individual is mentioned more than once. Thus, for example, "$Q_1a_1 \cdot Q_1a_2 \cdot Q_2a_4 \cdot Q_4a_6$" is a perfect description.

IA. Now consider first the case that H is a molecular sentence. Then H can always be transformed into a disjunction of perfect descriptions, all of which contain exactly the same individual

---

[9] In the case of a finite total population, the application of the simple product rule presupposes that the objects constituting a sample are taken from the urn one at a time and that each of them is replaced into the urn before the next one is drawn. In order to avoid complications of this sort, we assume the population to be infinite.

constants.* We omit the elementary but somewhat lengthy proof of this theorem here and rather illustrate it by an example. Let L contain exactly two primitive predicates, "P_1" and "P_2," and let $H_5 = "P_1a_1 \cdot P_2a_2"$; then H_5 can readily be expanded into the following expression:

$$"[(P_1a_1 \cdot P_2a_1) \vee (P_1a_1 \cdot {\sim}P_2a_1)] \cdot [(P_1a_2 \cdot P_2a_2) \vee ({\sim}P_1a_2 \cdot P_2a_2)]",$$

which in turn is equivalent to "$(Q_1a_1 \vee Q_2a_1) \cdot (Q_1a_2 \vee Q_3a_2)$"; and this can be transformed into the following disjunction of perfect descriptions:

$$"(Q_1a_1 \cdot Q_1a_2) \vee (Q_1a_1 \cdot Q_3a_2) \vee (Q_2a_1 \cdot Q_1a_2) \vee (Q_2a_1 \cdot Q_3a_2)".$$

Once H has thus been transformed, the determination of $pr(H, T, \Delta)$ follows simply the following two rules, which were illustrated above: (a) The probability of H with respect to T and Δ is the sum of the probabilities of the perfect descriptions whose disjunction is equivalent to H; (b) the probability of a perfect description with respect to T and Δ is the product of the relative frequencies assigned by Δ to the Q-expressions occurring in the perfect description. Thus, if $\Delta = \{q_1, q_2, q_3, q_4\}$, then $pr(H_5, T, \Delta) = q_1^2 + q_1q_3 + q_2q_1 + q_2q_3$.

IB. If H is a general sentence, then two cases have to be distinguished:

a.   If the number N of all objects to which L refers is finite, and if all of them have names, then H can obviously be transformed into a molecular sentence.† Thus, for example, the hypothesis "$(x)(P_1x \supset P_2x) \cdot (Ey)P_3y$" is equivalent to the following molecular sentence, which will also be called the *molecular development of* H *for the class of individuals* $\{a_1, a_2, \ldots, a_N\}$ – or, briefly, D_N (H): "$(P_1a_1 \supset P_2a_1) \cdot (P_1a_2 \supset P_2a_2) \ldots (P_1a_N \supset P_2a_N) \cdot (P_3a_1 \vee P_3a_2 \vee \ldots \vee P_3a_N)$." Now we simply define $pr(H, T, \Delta)$ as $pr(D_N(H), T, \Delta)$; and the latter magnitude can be determined according to the rules laid down in IA.

---

* Editor's note: This last qualification was omitted in the 1945 version.
† Editor's note: Quine pointed out the need for the requirement that all have names, which was missing in the 1945 version. It is needed in view of the bracketed material in the second paragraph of sec. 7.

b. If the class of all individuals is denumerably infinite and ordered in a sequence $a_1, a_2, a_3, \ldots$ – and this is the case with which we are principally concerned – then we define $pr(H, T, \Delta)$ as the limit, for indefinitely increasing N, of pr $(D_N(H), T, \Delta)$. It can be shown that this limit exists in all cases. (In particular, we note that when H is a general sentence containing no individual constants, the limit in question is either 0 or 1).

II. We now turn to the concept "$pr(H, E, \Delta)$," which refers to those cases where, besides the distribution Δ, some additional information E is given. To illustrate this case by means of the urn analogue and by reference to L_W, let again $\Delta_1 = \{\frac{1}{2}, \frac{1}{3}, 0, \frac{1}{6}\}$, and let $H_6 = $ "~Round a_5". Then $pr(H_6, T, \Delta_1) = \frac{1}{3} + \frac{1}{6} = \frac{1}{2}$. Now suppose that we are given the additional information $E_1 = $ "Blue a_5". In light of the thus enlarged total information, H_6 will acquire a different probability. Since, according to E_1, a_5 is blue, and since, according to Δ_1, the frequency of the nonround objects among the blue ones is $\frac{1}{3} \div (\frac{1}{2} + \frac{1}{3}) = \frac{2}{5}$, we shall set $pr(H_6, E_1, \Delta_1) = \frac{2}{5}$.

A completely general definition of $pr(H, E, \Delta)$ can be given in terms of the narrower concept "$pr(H, T, \Delta)$":

$$7.1 \qquad pr(H, E, \Delta) = \frac{pr(H \cdot E, T, \Delta)}{pr(E, T, \Delta)}$$

This definition presupposes that $pr(E, T, \Delta) \neq 0$; when this condition is not satisfied, $pr(H, E, \Delta)$ will not be defined.

This definition is suggested by the following consideration: We wish $pr(H, E, \Delta)$, for any fixed Δ, to satisfy the standard principles of probability theory,[10] including the general multiplication principle. Now the latter demands that

$$pr(E \cdot H, T, \Delta) = pr(E, T, \Delta) \cdot pr(H, E \cdot T, \Delta)$$

In view of the fact that $E \cdot H$ is logically equivalent to $H \cdot E$ and that $E \cdot T$ is logically equivalent to E, this leads to 7.1.

---

[10] These are stated in section 10 of the present article.

It can be proved that the concept thus defined satisfies all the customary postulates of probability theory.[11]

8. Optimum Distributions Relative to Given Evidence. Our problem of defining $dc(H, E)$ could now readily be solved if it were generally possible to infer from the given evidence E the frequency distribution Δ characteristic of the L-cells in the language under consideration, for we could then simply identify $dc(H, E)$ with $pr(H, E, \Delta)$. Unfortunately, however, no evidence sentence that is expressible in L can be strong enough to permit such an inference. Nonetheless, a closely related but somewhat weaker procedure is indeed available for the definition of $dc(H, E)$. This procedure is based on the fact that while a given E does not uniquely determine a fixed Δ, it may confer different degrees of likelihood – in a sense presently to be explained – upon the different possible distributions. Under favorable circumstances it may even be possible to characterize one particular distribution, Δ_E, as the one which is most likely on the basis of E; and in this case, $dc(H, E)$ might be defined as $pr(H, E, \Delta_E)$. We shall eventually extend this idea to the case where E does not uniquely determine just one most likely distribution; but before going into the details of this method, which will be done in the subsequent section, we have first to clarify the idea of likelihood referred to in the preceding discussion.

Let us illustrate the essential points by reference to L_W and the urn analogy. Suppose that E_1 is a report asserting that among 12 objects selected at random, 6 were blue and round, 4 blue and not round, and 2 neither blue nor round. If no additional information is available, we would say that in the light of the given evidence, $\Delta_1 = \{\frac{1}{2}, \frac{1}{3} \, 0, \frac{1}{6}\}$ is more likely than, say, $\Delta_2 = \{\frac{4}{9}, \frac{2}{9}, 0, \frac{3}{9}\}$ and that the latter is more likely than, say, $\Delta_3 = \{\frac{1}{10}, \frac{5}{10}, \frac{2}{10}, \frac{2}{10}\}$. Precisely how is the meaning of "more likely" to be construed here? It was shown in the preceding section that on the basis of any given frequency distribution Δ – and in the absence of any further infor-

---

[11] This probability concept was developed by Olaf Helmer; a detailed exposition of the theory of this concept is included in the article by Helmer and Oppenheim mentioned in footnote 1.

mation – it is possible to assign to every sentence S of L a definite probability pr (S, T, Δ). Here, S may be a hypothesis being tested or any other sentence in L. In particular, we may consider the case where S is our given evidence sentence E – that is, we may ask: What is the probability $pr(E, T, \Delta)$ which E would possess on the basis of a certain hypothetical distribution Δ and in the absence of any other information? If E is made more probable, in this sense, by a certain distribution Δ_1 than by another distribution Δ_2, then we shall say that Δ_1 has a *greater likelihood* relative to E than does Δ_2.

Illustration: In our last example, we have

$$pr(E_1, T, \Delta_1) = \left(\frac{1}{2}\right)^6 \cdot \left(\frac{1}{3}\right)^4 \cdot \left(\frac{1}{6}\right)^2$$

$$pr(E_1, T, \Delta_2) = \left(\frac{4}{9}\right)^6 \cdot \left(\frac{2}{9}\right)^4 \cdot \left(\frac{3}{9}\right)^2$$

$$pr(E_1, T, \Delta_3) = \left(\frac{1}{10}\right)^6 \cdot \left(\frac{5}{10}\right)^4 \cdot \left(\frac{2}{10}\right)^2$$

and, indeed, as can readily be verified, we have here

$$pr(E_1, T, \Delta_1) > pr(E_1, T, \Delta_2) > pr(E_1, T, \Delta_3),$$

in accordance with our earlier judgment as to the order of likelihoods involved.

Relative to some given evidence E, therefore, the infinitely many theoretically possible frequency distributions fall into a definite order of likelihood. By an *optimum distribution relative to E*, we shall understand a distribution Δ such that the probability $pr(E, T, \Delta)$, which Δ confers upon E, is not exceeded by the probability that any other distribution would assign to E.[12] Now it

---

[12] An alternative to this approach would be to determine, by means of Bayes' theorem, that distribution upon which E confers the greatest probability (in contradistinction to our question for that distribution which confers upon E the maximum probability). But this approach presupposes – to state it first by reference to the urn analogy – an infinity of urns, each with a different frequency distribution; and to each urn U, there would have to be assigned a definite a priori probability for the sample to be taken from U. Applied to our problem, this method would involve reference to an infinity of possible states

cannot be expected that every possible E determines exactly one optimum distribution: There may be several distributions, each one of which would give to E the same maximum probability. Thus, for example – to mention just one simple case – the probability of the evidence sentence "Blue a_1" in L_W will clearly be maximized by any distribution which makes the frequency of the blue objects equal to 1 (i.e., by any distribution of the form $\{q_1, 1 - q_1, 0, 0\}$, where q_1 may have any arbitrary value between 0 and 1 inclusive). It can be shown, however, that every E determines at least one optimum distribution. If there are several of them, then, of course, they all will confer the same probability upon E. We shall use the symbol "Δ_E" to refer to the optimum distribution or distributions relative to E; Δ_E is, therefore, a generally plurivalued function of the evidence E.

The determination of Δ_E for a given E is a mathematical problem whose treatment will be discussed here only in outline. Consider again the model language L_W and the four cells Q_1, Q_2, Q_3, Q_4 determined by it. Let a specific evidence sentence E be given. To find Δ_E, consider the general case of a hypothetical distribution $\Delta = \{q_1, q_2, q_3, q_4\}$, where the four components of Δ are parameters satisfying the conditions

8.1 $$0 \leqq q_s \leqq 1 \quad (s = 1, 2, 3, 4)$$

8.2 $$q_1 + q_2 + q_3 + q_4 = 1$$

The probability $pr(E, T, \Delta)$, which Δ confers upon E, will be a function $f(q_1, q_2, q_3, q_4)$ of the parameters, as is illustrated at the end of IA in section 6. Δ_E can now be found by determining those values of the parameters which satisfy 8.1 and 8.2 and for which $f(q_1, q_2, q_3, q_4)$ assumes an absolute maximum. These values are found by partial differentiation of the function f. By equating the partial derivatives to 0, we find that a system of simultaneous equations is obtained whose solution (or solutions) yield the value (or values) of Δ_E for the given evidence E. Explicit formulae for the solution of such systems of equations will be available

of the world, to each one of which there would have to be attached a certain a priori probability of being realized; and for such a "lottery of states of the world," as it were, it seems very difficult to find an empiricist interpretation.

only in special cases; but in many other cases, methods of computation can be indicated which will at least approximate the solutions. We mention here only one result of particular importance:

8.3   If E is a perfect description – as, for example, "$Q_1a_1 \cdot Q_1a_2 \cdot Q_1a_3 \cdot Q_2a_4 \cdot Q_2a_5 \cdot Q_3a_6$" in L_W – then Δ_E is unique, and its components are simply the relative frequencies with which the cells are represented in E – in our example, $\Delta_E = \{\frac{1}{2}, \frac{1}{3}, \frac{1}{6}, 0\}$.

The method used here to characterize optimum distributions goes back to a procedure introduced by R. A. Fisher as the *maximum likelihood method*.[13] We shall consider later the general character of our procedure; first we turn to the definition of dc in terms of the concept of optimum distribution.

9. Definition of dc(H, E).   In accordance with the program outlined in the beginning of the preceding section, we now define

9.1 $$dc(H,E) = pr(H,E,\Delta_E)$$

This definition embodies an empiricist reconstruction of the concept of degree of confirmation: On the basis of the given evidence E, we infer the optimum distribution (or distributions) Δ_E and then assign to H, as its degree of confirmation, the probability which H possesses relative to E according to Δ_E.

As can be seen from 7.1, the definition 9.1 determines $dc(H, E)$ in all cases where $pr(E, T, \Delta_E) \neq 0$. Now it can be shown that this condition is satisfied if and only if E is logically consistent, so that, by 9.1, $dc(H, E)$ is defined for every noncontradictory E.

It should be noted, however, that, since Δ_E is not necessarily single-valued, $dc(H, E)$ may have more than one value.[14]

---

[13] Compare R. A. Fisher, "The Mathematical Foundations of Theoretical Statistics," *Philosophical Transactions of the Royal Society of London* 222 (1922), 309–68. Also see M. G. Kendall, "On the Method of Maximum Likelihood," *Journal of the Royal Statistical Society* 103 (1940), 388–99, and the same author's work, *Advanced Theory of Statistics* (London: Criffin, 1943).

[14] The symbol "$dc(H, E)$" is, therefore, used here in a similar manner as, say, "$\sqrt{x}$" in mathematics; both represent functions which are not generally single-

Thus, for example, when $H = "P_1a"$ and $E = "P_2b"$, $dc(H, E)$ turns out to have as its values all the real numbers between 0 and 1 inclusive. This is quite sensible in view of the fact that the given E is entirely irrelevant for the assertion made by H; E, therefore, can impose no restrictions at all upon the range of the logically possible values of the degree to which H may be confirmed.

However, $dc(H, E)$ can be shown to be single-valued in large classes of cases. These include, in particular, the cases where E is a perfect description, as can readily be seen from theorem 8.3. Also, it can be shown that in all cases of the kind characterized in 5.3, our definition leads to a unique value of $dc(H, E)$ and that this value is the relative frequency stipulated in 5.3.

We shall now analyze in some detail a special example which incidentally shows that dc can be single-valued in cases other than those just mentioned. Let L contain just one primitive predicate, $"P"$, and let

$$E = "(Pa_1 \cdot Pa_2 \cdot {\sim}Pa_3) \vee (Pa_1 \cdot {\sim}Pa_2 \cdot {\sim}Pa_3) \vee ({\sim}Pa_1 \cdot {\sim}Pa_2 \cdot Pa_3)"$$

and

$$H_1 = "Pa_4", H_2 = "Pa_1"$$

In order to determine Δ_E, we have to find that $\Delta = \{q, 1 - q\}$, which maximizes the magnitude

9.2 $$pr(E, T, \Delta) = q^2(1 - q) + q(1 - q)^2 + (1 - q)^3 q$$
$$= q^3 - 3q^2 + 2q$$

By equating the derivative of this function to 0 and solving for q, we obtain

9.3 $$q = \frac{3 - \sqrt{3}}{3}$$

valued. An alternative would be to stipulate that $dc(H, E)$ is to equal $pr(H, E, \Delta_E)$ in those cases where the latter function is single-valued, and that in all other cases, $dc(H, E)$ is to remain undefined. A third possibility would be to define $dc(H, E)$ as the smallest value of $pr(H, E, \Delta_E)$, for of two hypotheses tested by means of the same evidence, that one will be considered more reliable for which that smallest value is greater. This definition, however, has a certain disadvantage, which is explained in footnote 17.

and hence

9.4
$$\Delta_E = \left\{ \frac{3-\sqrt{3}}{3}, \frac{\sqrt{3}}{3} \right\}$$

Substituting from 9.3 in 9.2 yields

9.5
$$pr(E,T,\Delta_E) = \frac{2\sqrt{3}}{9}$$

We similarly compute

9.6
$$pr(H_1 \cdot E,T,\Delta_E) = \frac{2\sqrt{3}}{9} \cdot \frac{3-\sqrt{3}}{3}$$

Hence,

9.7
$$dc(H_1,E) = \frac{pr(H_1 \cdot E,T,\Delta_E)}{pr(E,T,\Delta_E)} = \frac{3-\sqrt{3}}{3}$$

As for H_2, we note that

9.8
$$H_2 \cdot E = "(Pa_1 \cdot Pa_2 \cdot {\sim}Pa_3) \vee (Pa_1 \cdot {\sim}Pa_2 \cdot {\sim}Pa_3)"$$

Hence,

9.8
$$pr(H_2 \cdot E,T,\Delta_E) = q^2(1-q) + q(1-q)^2$$
$$= q(1-q) = \frac{\sqrt{3}-1}{3}$$

and finally

9.9
$$dc(H_2,E) = \frac{pr(H_2 \cdot E,T,\Delta_E)}{pr(E,T,\Delta_E)} = \frac{3-\sqrt{3}}{2}$$

After having considered some examples involving nongeneral hypotheses, we now turn to the case of hypotheses in the form of general sentences. Let us assume, for example, that L contains again only one primitive predicate, "P", and let H = "$(x)Px$", E_1 = "Pa_1", E_2 = "$Pa_1 \cdot Pa_2 \cdot \ldots Pa_t$", E_3 = "${\sim}Pa_1 \cdot Pa_2 \cdot Pa_3 \cdot \ldots Pa_t$". To compute the values of dc for these cases, we note first that, as can readily be shown, conditions 5.1 and 5.2 are satisfied by dc as defined in 9.1 and that, as a consequence, $dc(H, E) = 0$ whenever H contradicts E.

Now, if again we assume the class of all objects to be infinite, we obtain

$$dc(H,E_1) = dc(H,E_2) = \lim_{N \to \infty} 1^N = 1; \quad dc(H,E_3) = 0,$$

no matter now large *t* may be. The last value appears perfectly reasonable: Since E_3 contains one conjunctive term which contradicts H, E_3 itself contradicts H and thus disconfirms it to the highest degree that is theoretically possible. The value 1 in the first two cases, however, might seem counterintuitive for two reasons: First, it seems strange that it should make no difference for the value of $dc(H, E)$ how many confirming instances for H are included in E – as long as E contains no disconfirming evidence; and second, it is surprising that even one single confirming case for H should confirm the hypothesis H – which virtually covers an infinity of such cases – to the maximum extent. The significance of these results might become clearer if we distinguish between the retrospective and the prospective aspects of what has sometimes been called the probability – and what we call the degree of confirmation – of a universal hypothesis. Taken retrospectively, the magnitude in question is to characterize the extent to which H is confirmed "by past experience" (i.e., by the given evidence E); taken prospectively, it is to constitute, as it were, a measure of the warranted assertability of the hypothesis – or of the rational belief to be placed in its validity in instances which have as yet not been examined. Now clearly, in our illustration, H is confirmed to the fullest possible extent by E_1 as well as by E_2. In both cases it is satisfied in 100% of the instances mentioned by E. As to the prospective aspect, it is simply an inductivist attitude which directs us to assign the *dc* 1^N to the hypothesis that the next N instances will conform to the hypothesis, and finally, the limit of 1^N, for indefinitely increasing N, to the hypothesis itself (i.e., to the assumption that *all* objects conform to it).

10. Probability and Degree of Confirmation. Might $dc(H, E)$ as well be called the probability of the hypothesis H relative to the evidence E? Partly, of course, that is a matter of arbitrary terminological decision. However, the concept of probability has come to be used with reference to magnitudes that satisfy certain conditions, which, for brevity, will be called here the postulates of general probability theory.[15] We shall summarize them here in a form adapted from Janina Hosiasson-Lindenbaum's article "On confirmation."[16]

The probability of H relative to E, or, briefly, $p(H, E)$, is a single-valued function of two sentences, the second of which is noncontradictory. This function satisfies the following conditions:

10.1   If H is a consequence of E, then $p(H, E) = 1$.

10.2   If E implies that H_1 and H_2 cannot both be true, then

$$p(H_1 \vee H_2, E) = p(H_1, E) + p(H_2, E)$$

(Special addition principle of probability theory)

10.3 $p(H_1 \cdot H_2, E) = p(H_1, E) \cdot p(H_2, H_1 \cdot E)$

(General multiplication principle of probability theory)

10.4   If E_1 and E_2 are logically equivalent, then

$$p(H, E_1) = p(H, E_2)$$

The concept "$pr(H, E, \Delta)$" can be shown to satisfy, for any fixed Δ, all of these conditions. But the concept "$dc(H, E)$", which is defined by reference to it, does not. For, first, as we saw, $dc(H, E)$ is not always a single-valued function of H and E. As to the four postulates listed above, the following can be shown: The first, second, and fourth postulates are generally satisfied by dc provided that when dc is plurivalued, "corresponding values" – that is, values obtained from the same Δ_E – are substituted in the formulae. The third postulate, however, is not generally satisfied. The reason for this becomes clear when in 10.3, "dc" is replaced by its definiens. Then the left-hand side turns into "$pr(H_1 \cdot H_2, E, \Delta_E)$", and as "$pr$" satisfies the general multiplication principle, we may transform the last expression into "$pr(H_1, E, \Delta_E) \cdot pr(H_2, H_1 \cdot E, \Delta_E)$"; but the right-hand side of 10.3 transforms into "$pr(H_1, E,$

---

[15] On this point, compare also section 3 of Professor Carnap's article (see footnote 1).

[16] *The Journal of Symbolic Logic* 5 (1940), pp. 133–48.

$\Delta_E) \cdot pr(H_2, H_1 \cdot E, \Delta_{H_1 \cdot E})$". Clearly, the second factors in these two expressions cannot generally be expected to be equal. However, the following restricted version of 10.3 is generally satisfied:

10.3' "Corresponding values" of $dc(H, E)$ satisfy 10.3 in particular if the two following conditions are satisfied:
a. H_1 and H_2 have no individual constants in common,
b. At least one of the hypotheses H_1, H_2 has no individual constants in common with E.[17]

In view of the fact that dc as defined above does not satisfy all of the postulates of probability theory, we prefer not to call dc a probability.[17a]

Finally, it may be of interest to compare our way of defining dc with another method, which makes use of the concept of measure of a sentence. Briefly, this method consists of assigning, by means of some general rule, a measure $m(S)$ to every sentence S in L in such a manner that the following conditions are satisfied:

10.5.1   For every S, $0 \leqq m(S) \leqq 1$.

10.5.2   If S_1, S_2 are logically equivalent, then $m(S_1) = m(S_2)$.

10.5.3   If S_1, S_2 are logically incompatible, then

$$m(S_1 \vee S_2) = m(S_1) + m(S_2)$$

10.5.4   For any analytic sentence T, $m(T) = 1$.

---

[17] In footnote 14, two alternatives to our definition of dc were mentioned. It can be shown that the concept determined by the first of these satisfies without exception the requirements 10.1, 10.2, 10.3', and 10.4, whereas the concept introduced by the second alternative does not. Thus, for example, if $H = $ "P_1a_1", $E = $ "P_2a_2", then the values of $pr(H, E, \Delta_E)$ are all the real numbers from 0 to 1 inclusive, so that the smallest value is 0. The same is true of $pr(\sim H, E, \Delta_E)$; hence these two smallest values violate the principle 5.1 and thus indirectly the postulates 10.1 and 10.2, of which 5.1 can be shown to be a consequence.

[17a] The alternative term *likelihood* which suggests itself is inexpedient also, as it has already been introduced into theoretical statistics with a different meaning (cf. section 8 earlier). If a term customarily associated with "probability" should be desired, then "expectancy" might be taken into consideration.

The degree of confirmation of a hypothesis H with respect to a noncontradictory evidence sentence E is then defined as $\dfrac{m(H \cdot E)}{m(E)}$. The four stipulations 10.5 leave room for an infinite variety of possible measure functions; the choice of a particular function will be determined by the adequacy of the concept of degree of confirmation which is definable in terms of it.[18] Our concept "$dc(H, E)$" can be introduced in a formally similar manner as follows: Instead of assigning to each sentence of L once and for all an apriori measure, as it is done in the method just described, we give to the sentences of L measures which depend on the given empirical evidence E. The E-measure of a sentence S in L can be defined thus:

10.6 $$m_E(S) = pr(S, T, \Delta_E)$$

In terms of this magnitude, we can express $dc(H, E)$ as follows:

10.7 $$dc(H, E) = \frac{m_E(H \cdot E)}{m_E(E)},$$

for by virtue of 9.1, 7.1, and 10.6,

$$dc(H, E) = pr(H, E, \Delta_E) = \frac{pr(H \cdot E, T, \Delta_E)}{pr(E, T, \Delta_E)} = \frac{m_E(H \cdot E)}{m_E(E)},$$

[As was pointed out in connection with 9.1, $m_E(E) = pr(E, T, \Delta_E)$ equals 0 only when E is contradictory; in this case, $dc(H, E)$ is not defined.]

11. *Concluding Remarks.* The concept of dc as it has been defined here is a purely logical concept in the following sense:

---

[18] The method characterized above is illustrated by a definition of probability which F. Waismann ("Logische Analyse des Wahrscheinlichkeitsbegriffs," *Erkenntnis*, vol. 1, pp. 228–48) has outlined following a suggestion made in L. Wittgenstein's *Tractatus Logico-Philosophicus* (New York and London: Kegan Paul, 1922). Also, the regular c-functions introduced in Professor Carnap's article, "On inductive logic" (see footnote 1), exemplify this way of defining dc. In that article, some special choices for the measure function m are presented and examined as to their suitability for the establishment of an adequate definition of the concept of degree of confirmation.

Given two sentences H and E in L, $dc(H, E)$ is completely determined by the formal, or syntactical, structure of the two sentences alone; and apart from possible mathematical complications, its value can be found by an analysis of that structure and the application of certain purely deductive mathematical techniques. Nevertheless, the proposed concept is empiricist and not "aprioristic" in character, for the degree of confirmation assigned to H is determined, generally speaking, by reference to relative frequencies derived from the evidence sentence E. With reference to the alternative definition 10.6, the matter can be stated as follows: dc is defined in terms of the concept of measure of a sentence; but whereas in an aprioristic theory the measure of a sentence is determined once and for all on the basis of a mere analysis of its logical structure, the measure used in 10.6 is empiricist insofar as its determination requires reference not only to the structure of the sentence but also to the given empirical evidence E.

The method employed to determine $dc(H, E)$ consists essentially of two steps: First, by means of the maximum likelihood principle, a hypothetical assumption is formed, on the basis of E, as to the frequency distribution for the L-cells; second, on the basis of the hypothetical distribution thus assumed, a probability is assigned to H relative to E. The rationale of this procedure is perhaps best exhibited by reference to a simple model case. Suppose that we are given a die about whose homogeneity and symmetry nothing is known. We have an opportunity to roll the die twenty times and are then to lay a bet on the hypothesis H that both the 21st and the 22nd throw will yield a six. The maximum likelihood principle would direct us, in this particularly simple case, to record, in a report E, the occurrence or nonoccurrence of a six as the result of each of the first twenty throws and then to form a hypothesis as to the limit of the relative frequency with which throws with the given die will yield a six. This limit is to be chosen in such a way that relative to it, the distribution of the results found in E has a maximum probability. In the simple case under consideration, this means that we have to set the limit equal to the relative frequency with which the result six is reported in E. Let this be $\frac{1}{10}$; then the distribution $\Delta_E = \{\frac{1}{10}, \frac{9}{10}\}$

for the cells corresponding to the results six and non-six is the optimum distribution, and on the basis of it, $dc(H, E)$ becomes $\frac{1}{100}$. This value would be the basis for determining the rates of a fair bet on H, in light of E.

In this special case, which is covered by rule 5.3, the procedure dictated by the maximum likelihood principle clearly coincides with that which a "rational gambler" would use – and which is also used in statistical investigations of various kinds. This procedure reflects an assumption which might be called the statistical version of the principle of induction and which, stated in very crude terms, implies that relative frequencies observed "in the past" (i.e., in the instances so far examined) will remain fairly stable "in the future" (i.e., in those instances which have not as yet been examined, no matter whether they belong to the past or to the future). The maximum likelihood principle in the form in which it has been used here for the general definition of dc is but an extension of this same idea to cases more complex than those covered by rule 5.3; and we may say that it represents a generalization and rational reconstruction of the statistical version of the principle of induction.

The theory obtained by our procedure provides criteria which establish, so to speak, a fair rate of betting on a specified hypothesis on the basis of given data. (In many cases, as we saw, dc will be single-valued, and the betting rate will, therefore, be uniquely determined; in other cases, where the evidence is insufficient in a certain sense, dc will have several values, and, then, the smallest of these might be used to establish a betting rate.) The decisions, however, which a gambler has to make concern not only the betting rate but also the amount he is going to risk; and while the rate is determined, generally speaking, by the relative frequency in the past of the event on which he wishes to bet, the gambler's stake will be determined by different factors, such as, for example, the size of the sample which represents the evidence. Analogously, the concept of degree of confirmation, as it has been defined in the present article, refers only to one among several factors which enter into an objective appraisal of the soundness or reliability of an empirical hypothesis. The remaining factors include, among others, the number of tested instances which are mentioned in E, and the variety of those instances.[19] Our theory of confirmation is intended to account exclusively for the first of these various aspects of the evaluation of a hypothesis by means of relevant evidence – that aspect which is analogous to the betting rate in the preceding example.

The theory of confirmation which has been outlined in this article cannot claim to be more than a first contribution to the exploration of a field in which systematic logical research is only beginning. Among various problems which are suggested by the present study, we should like to point out a few which seem to deserve special attention in future research:

1. The next step in the development of the theory of confirmation would be the extension of the definition of dc to the entire lower functional calculus and possibly even to the higher functional calculus.

2. In section 1, we distinguished the metrical concept of degree of confirmation from the classificatory concepts of confirming and disconfirming evidence for a given hypothesis. In this connection, the question arises whether the meaning of the expressions "E is confirming evidence for H" and "E is disconfirming evidence for H" is adequately definable in terms of $dc(H, E)$.

3. In the practice of scientific research, observation reports are not all considered equally reliable; rather, their reliability will depend on certain characteristics of the observer, on the instruments of observation used, and on the circumstances under which the observation took place. Also, when general sentences are included in the evidence E, these might be said to have different degrees of reliability (which, for example, might be determined on the basis of their dc relative to all the relevant evidence known at the time). We might try to reflect this aspect of scientific testing by assuming in our theory that each evidence sentence is assigned a numerical

[19] Compare Ernest Nagel, "Principles of the Theory of Probability," *International Encyclopedia of Unified Science* 1, 6 (1939), 68–71. Also, see section 15 of Professor Carnap's paper (see footnote 1), which contains a discussion of this point.

"weight" whose value is a real number between 0 and 1, inclusive. The problem then arises of defining $dc(H, E)$ in a manner which takes into consideration those weights attached to the evidence. The generalized definition here called for should comprehend, as one special case, our definition 9.1 (or another adequate definition of this kind); for the latter rests, as it were, on the tacit assumption that the weight of the given evidence is always 1.

Methodology

9   [3]* "Analyse logique de la psychologie," *Revue de Synthese* 10 (1935), 27–42. Wilfrid Sellars's translation, "The Logical Analysis of Psychology," appeared in Herbert Feigl and Wilfrid Sellars, eds., *Readings in Philosophical Analysis* (New York: Appleton-Century-Crofts, 1949), pp. 373–84.

10 [92] "Schlick und Neurath: Fundierung *vs.* Kohärenz in der wissenschaftlichen Erkenntnis," *Grazer philosophische Studien*, vol. 16/17 (1983), 1–18; Christian Piller, trans., "Schlick and Neurath: Foundation *vs.* Coherence in Scientific Knowledge."

11 [102] "On the Cognitive Status and the Rationale of Scientific Methodology," <u>Poetics Today</u> 9 (1988), 5–27. [/] Hempel notes that Part I <u>is an annotated</u> and slightly revised version of [95] and that Part II contains passages from [97].

12 [100] "Provisoes: A Problem Concerning the Inferential Function of Scientific Theories," in Adolf Grunbaum and Wesley Salmon, eds., *The Limitations of Deductivism* (Berkeley: University of California Press, 1988), pp. 19–36. Also in *Erkenntnis* 28 (1988), 147–64.

Essay 9 is a period piece from <u>the early heyday of physicalism,</u> <u>when it could seem plausible both that any meaningful statement</u> <u>must be translatable into a sentence in the language of physics</u> <u>and that the possibility of such translation</u> is a purely logical matter. In it, Hempel's proposed dissolution of the mind-body problem was a version of the identity theory according to which it is a matter of logic that "Otto has a toothache" will be equipol-

---

* The numbers in brackets represent the publication number for each chapter here in this list. (See the section entitled "C. G. Hempel's Publications" beginning on p. 305.)

lent to some physical sentence. Which physical sentence? The answer is an empirical matter.

Essay 10 is a recollection in tranquillity of the heated protocol sentence debate (essays 1–3), revisited here with Kuhn and Quine very much in mind.

Essay 11 is Hempel's late view of methodology in full flower. Here, as in some other late essays, the curtain rises on Carnap or Popper (stage right) and Neurath or Kuhn or Feyerabend (stage left) in the roles of the aprioristic and naturalistic poles of the methodological globe. As the play develops, we are made to see that the real Carnap and Neurath (and the others) are not to be found at either pole.

Essay 12 is a remarkable further development of the late view, in which Hempel presents a new impediment to the logical empiricist vision of a scientific theory as a deductive bridge from initial conditions to predictions, both of which are stated in purely observational terms. Here he argues that such deductions are possible only under "provisoes" – essential but generally tacit presuppositions that cannot be stated in purely observational terms – and he argues that the need for such provisoes makes nonsense of various key concepts of logical empiricism (e.g., the concept of the empirical content of a scientific theory).

Chapter 9
The Logical Analysis of Psychology

(1935)

I

One of the most important and most discussed problems of contemporary philosophy is that of determining how psychology should be characterized in the theory of science. This problem, overflowing the limits of epistemological analysis and leading to heated controversy in metaphysics itself, is brought to a focus by this familiar disjunction: "Is psychology a natural science, or is it one of the sciences of mind and culture (*Geisteswissenchaften*)?"

The present article attempts to sketch the general lines of a new analysis of psychology, one which makes use of rigorous logical tools and which has made possible decisive advances toward the solution of the above problem.[1] This analysis was successfully

---

[1] I now (1947) consider the type of physicalism outlined in this paper as too restrictive; the thesis that all statements of empirical science are *translatable*, without loss of theoretical content, into the language of physics should be replaced by the weaker assertion that all statements of empirical science are *reducible* to sentences in the language of physics, in the sense that for every empirical hypothesis, including, of course, those of psychology, it is possible to formulate certain test conditions in terms of physical concepts which refer to more or less directly observable physical attributes. But those test conditions are not asserted to exhaust the theoretical content of the given hypothesis in all cases.

For a more detailed development of this thesis, compare R. Carnap, "Logical Foundations of the Unity of Science," in *International Encyclopedia of*

undertaken by the Vienna Circle (*Wiener Kreis*), the members of which (M. Schlick, R. Carnap, Phillipp Frank, O. Neurath, F. Waismann, H. Feigl, etc.) have, during the past ten years, developed an extremely fruitful method for the epistemological examination and critique of the various sciences, based in part on the work of L. Wittgenstein.[2] We shall limit ourselves essentially to the examination of psychology as carried out by Carnap and Neurath.

The method characteristic of the studies of the Vienna Circle can be briefly defined as a *logical analysis of the language of science*. This method became possible only with the development of an extremely subtle logical apparatus which makes use, in particular, of all the formal procedures of modern logistics.[3] However, in the following account, which does not pretend to give more than a broad orientation, we shall limit ourselves to the aim of bringing out the general principles of this new method, without making use of strictly formal procedures.

II

Perhaps the best way to bring out the meaning and scope of the position of the Vienna Circle as it relates to psychology is to say that it is the exact antithesis of the current epistemological conviction that there is a fundamental difference between experimental psychology, as a natural science, and introspective psychology – in general, between the natural sciences as a whole and the sciences of mind and culture.[4] The common content of the

Unified Science, vol. I, N. 1. (Chicago: University of Chicago Press, 1938), pp. 42–62.

[2] *Tractatus Logico-Philosophics* (London: Kegan Paul, 1922).

[3] A recent presentation of logistics, based on the fundamental work of Whitehead and Russell, *Principia Mathematica*, is to be found in R. Carnap, *Abriss der Logistik* (Vienna: Julius Springer, 1929). (This book was volume II of the series, *Schriften zur Wissenschaftlichen Weltauffassung*.) Carnap's article includes an extensive bibliography, as well as references to other logistics systems.

[4] The following are some of the principal publications of the Vienna Circle on the nature of psychology as a science: R. Carnap, *Scheinprobleme in der Philosophie: Das Fremdpsychische und der Realismusstreit* (Berlin-Schlachtensee:

widely different formulae which are generally used to express this contention, which we reject, can be set down as follows: Apart from certain aspects clearly related to physiology, psychology is radically different, both as to subject-matter and as to method, from physics in the broad sense of the term. In particular, it is impossible to deal adequately with the subject-matter of psychology by means of physical methods. The subject-matter of physics includes such concepts as mass, wave length, temperature, field intensity, etc. In developing these concepts, physics employs its distinctive method, which makes a combined use of description and causal explanation. Psychology, on the other hand, has for its subject-matter notions which are, in a broad sense, mental. They are *toto genere* different from the concepts of physics, and the appropriate method for dealing with them scientifically is that of sympathetic insight, called "introspection," a method which is peculiar to psychology.

One of the essential differences between the two kinds of subject-matter, it is believed, consists of the fact that the objects investigated by psychology – in contradistinction to physics – possess an intrinsic meaningfulness. Indeed, several proponents of this idea state that the distinctive method of psychology consists of "understanding the sense of significant structures" (*sinnvolle Gebilde verstehend zu erfassen*). Take, for example, the case of a man who speaks. Within the framework of physics, this process is considered to be completely explained once one has traced the movements which make up the utterance to their causes – that is to say, to certain physiological processes in the organism, and, in particular, to the central nervous system. But,

Weltkreis, 1928); Carnap, *Der logische Aufbau der Welt* (Berlin-Schlachtensee: Weltkreis, 1928; Hamburg: Felix Meiner, 1961); Rolf A. George, trans., *The Logical Structure of the World* (Berkeley: University of California Press, 1967); Carnap, "Die physikalische Sprache als Universalsprache der Wissenschaft," *Erkenntnis* 2 (1931), 432–65; Carnap, "Psychologie in physikalischer Sprache," *Erkenntnis* 3 (1932), 107–42; "Über Protokollsätze," *Erkenntnis* 3 (1932), 215–28; O. Neurath, "Protokollsätze," *Erkenntnis* 3 (1932), 204–14; Neurath, *Einheitswissenschaft und Psychologie* in the series *Einheitswissenschaft* (Vienna: Gerold, 1933), pp. 547–610. See also the publications mentioned in the notes below.

it is said, this does not even broach the psychological problem. The latter begins with an understanding of what was said and proceeds to integrate it into a wider context of meaning.

It is usually this latter idea which serves as a principle for the fundamental dichotomy that is introduced into the classification of the sciences. There is taken to be an *absolutely impassable gulf* between the *natural sciences*, which have a subject-matter devoid of sense, and the *sciences of mind and culture*, which have an intrinsically meaningful subject-matter, the appropriate methodological instrument for the scientific study of which is "insight into meaning."

III

The position in the theory of science which we have just sketched has been attacked from several different points of view.[5] As far as psychology is concerned, one of the principal countertheses is that formulated by behaviorism, a theory born in America shortly before the war. (In Russia, Pavlov has developed similar ideas.) Its principal methodological postulate is that a scientific psychology should limit itself to the study of the bodily behavior with which man and the animals respond to changes in their physical environment – every descriptive or explanatory step which makes use of such terms from introspective or "understanding" psychology as "feeling," "lived experience," "idea," "will," "intention," "goal," "disposition," or "repression" being proscribed as nonscientific.[6] We find in behaviorism, consequently, an attempt to construct a scientific psychology which

[5] P. Oppenheim, for example, in his book *Die Natuerliche Ordnung der Wissenschaften* (Jena, Germany: Fischer, 1926), opposes the view that there are fundamental differences between any of the different areas of science. On the analysis of "understanding," compare M. Schlick, "Erleben, Erkennen, Metaphysik," *Kantstudien*, 31 (1926), 146–58.

[6] For further details see the statement of one of the founders of behaviorism: J. B. Watson, *Behaviorism* (New York: People's Institute, 1925). Also see A. A. Roback, *Behaviorism and Psychology* (Cambridge: The University Bookstore, 1923); and A. P. Weiss, *A Theoretical Basis of Human Behavior*, 2nd ed. rev., (Columbus, OH: Adams, 1929). See also the work by Koehler cited in footnote 10 below.

would show by its success that even in psychology we have to do with purely physical processes and that, therefore, there can be no impassable barrier between psychology and physics. However, this manner of undertaking the critique of a scientific thesis is not completely satisfactory. It seems, indeed, that the soundness of the behavioristic thesis expounded above depends on the possibility of fulfilling the program of behavioristic psychology. But one cannot expect the question as to the scientific status of psychology to be settled by empirical research in psychology itself. To achieve this is rather an undertaking in epistemology. We turn, therefore, to the considerations advanced by members of the Vienna Circle concerning this problem.

IV

Before attacking the question as to whether the subject-matters of physics and psychology are essentially the same or different in nature, it is necessary first to clarify the very concept of the subject-matter of a science. The theoretical content of a science is to be found in propositions. It is necessary, therefore, to determine whether there is a fundamental difference between the propositions of psychology and those of physics. Let us, therefore, ask what it is which determines the content – one can equally well say the "meaning" – of a proposition. When, for example, do we know the meaning of the following statement: "Today at one o'clock, the temperature of such and such a place in the physics laboratory was 24.3° centigrade"? Clearly when, and only when, we know under what conditions we would characterize the statement as true, and under what circumstances we would characterize it as false. (Needless to say, it is not necessary to know whether or not the statement is true.) Thus, we understand the meaning of the above statement since we know that it is true when a tube of a certain kind, filled with mercury (in short, a thermometer with a centigrade scale) and placed at the indicated time at the location in question, exhibits a coincidence between the level of the mercury and the mark of the scale numbered 23.4. It is also true if in the same circumstances one can observe certain coincidences on another instrument called an "alcohol thermometer"; and, again, if a galvanometer connected to a thermopile shows a certain deviation when the thermopile is placed there at the indicated time. Finally, there is a long series of other possibilities which make the statement true, each of which is defined by a "physical test sentence," as we should like to call it. The statement itself clearly affirms nothing other than this: All these physical test sentences obtain. (However, one verifies only some of these physical test sentences and then "concludes by induction" that the others obtain as well.) The statement, therefore, is nothing but an abbreviated formulation of all these test sentences.

Before continuing the discussion, let us sum up this result as follows:

1. A proposition that specifies the temperature at a selected point in space-time can be "retranslated" without change of meaning into another proposition – doubtlessly longer – in which the word "temperature" no longer appears. This term functions solely as an abbreviation, making possible the concise and complete description of a state of affairs, the expression of which would otherwise be very complicated.

2. The example equally shows that *two propositions which differ in formulation* can nevertheless have the *same meaning*. A trivial example of a statement having the same meaning as the above would be: "Today at one o'clock, at such and such a location in the laboratory, the temperature was 19.44° Réaumur."*

As a matter of fact, the preceding considerations show – and let us set it down as another result – that *the meaning of a proposition is established by the conditions of its verification*. In particular, two differently formulated propositions have the same meaning or the same effective content when, and only when, they are both true or both false in the same conditions. Furthermore, a proposition for which one can indicate absolutely no conditions which would verify it, which is in principle incapable of confrontation with test conditions, is wholly devoid of content and without

---

* In the Réaumur scale, the ice point is 0 degrees and the steam point is 80 degrees. (Editor)

meaning. In such a case we have to do with a "pseudoproposition" – that is to say, a sequence of words correctly constructed from the point of view of grammar – but without content – rather than with a proposition properly speaking.[7]

In view of these considerations, our problem reduces to one concerning the difference between the circumstances which verify psychological propositions and those which verify the propositions of physics. Let us, therefore, examine a proposition which involves a psychological concept – for example, "Paul has a toothache." What is the specific content of this proposition – that is to say, what are the circumstances in which it would be verified? It will be sufficient to indicate some test sentences which describe these circumstances.

a. Paul weeps and makes gestures of such and such kinds.
b. At the question, "What is the matter?" Paul utters the words, "I have a toothache."
c. Closer examination reveals a decayed tooth with exposed pulp.
d. Paul's blood pressure, digestive processes, and the speed of his reactions show such and such changes.
e. Such and such processes occur in Paul's central nervous system.

This list could be expanded considerably, but it is already sufficient to bring out the fundamental and essential point – namely, that all the circumstances which verify this psychological proposition are expressed by physical test sentences. [This is true even of test sentence b, which merely expresses the fact that in specified physical conditions (the propagation of vibrations produced in the air by the enunciation of the words "What is the matter?"), there occurs in the body of the subject a certain physical process (speech behavior of such and such a kind).]

The proposition in question, which is about someone's "pain,"

---

[7] Space is lacking for a further discussion of the logical form of a test sentence (recently called "protocol-propositions" by Neurath and Carnap). On this question see Wittgenstein, *Tractatus Logico-Philosophicus*, as well as the articles by Neurath and Carnap which have appeared in *Erkenntnis* (above, footnote 4).

is, therefore, equally with that concerning the temperature, simply an abbreviated expression of the fact that all its test sentences are verified. (Here, also, one verifies only some of the test sentences and then infers by way of induction that the others obtain as well.) It can be retranslated without loss of content into a proposition which no longer involves the term "pain," but only physical concepts. Our analysis has consequently established that a certain proposition belonging to psychology has the same content as a proposition belonging to physics; and this result is in direct contradiction with the thesis that there is an impassable gulf between the statements of psychology and those of physics.

The above reasoning can be applied to *any psychological proposition*, even to those which concern, as is said, "deeper psychological strata" than that of our example. Thus, the assertion that Mr. Jones suffers from intense inferiority feelings of such and such kinds can be confirmed or falsified only by observing Mr. Jones' behavior in various circumstances. To this behavior belong all the bodily processes of Mr. Jones and, in particular, his gestures, the flushing and paling of his skin, his utterances, his blood pressure, the events that occur in his central nervous system, etc. In practice, when one wishes to test propositions concerning what are called the deeper layers of the psyche, one limits oneself to the observation of external bodily behavior – and, particularly, to speech movements aroused by certain physical stimuli (the asking of questions). But it is well known that experimental psychology has also developed techniques for making use of the subtler bodily states referred to above in order to confirm the psychological discoveries made by cruder methods. The statement concerning the inferiority feelings of Mr. Jones – whether true or false – means only this: Such and such happenings take place in Mr. Jones' body in such and such circumstances.

We shall call a proposition which can be translated without change of meaning into the language of physics a "physicalistic proposition," whereas we shall confine the expression "proposition of physics" to those statements which are already formulated in the terminology of physical science. (Since every statement is in respect of content equivalent – or, better, equipollent – to itself, every proposition of physics is also a physicalistic proposition.) The result of the preceding considerations can now be summed up as follows: *All psychological statements which are meaningful – that is to say, which are in principle verifiable – are translatable into propositions which involve not psychological concepts but only the concepts of physics. The propositions of psychology are consequently physicalistic propositions. Psychology is an integral part of physics.* If a distinction is drawn between psychology and the other areas of physics, it is only from the point of view of the practical aspects of research and the direction of interest, rather than as a matter of principle. This logical analysis, of which the result shows a certain affinity with the fundamental ideas of behaviorism, constitutes the physicalistic conception of psychology.

V

It is customary to raise against the above conception the following fundamental objection: The physical test sentences of which you speak are absolutely incapable of formulating the intrinsic nature of a mental process; they merely describe the physical *symptoms* from which one infers, by purely psychological methods – notably that of understanding – the presence of a certain mental process. But it is not difficult to see that the use of the method of understanding or of other psychological procedures is bound up with the existence of certain observable physical data concerning the subject undergoing examination. There is no psychological understanding that is not tied up physically in one way or another with the person to be understood. Let us add that, for example, in the case of the proposition about the inferiority complex, even the "introspective" psychologist, the psychologist who "understands," can confirm his conjecture only if the body of Mr. Jones, when placed in certain circumstances (most frequently, subjected to questioning), reacts in a specified manner (usually, by giving certain answers). Consequently, even if the proposition in question had to be arrived at, *discovered*, by "sympathetic understanding," the only *information* it gives us is nothing more nor less than the following: Under certain circumstances, certain specific events take place in the body of Mr. Jones. It is this which constitutes the meaning of the psychological statement.

The further objection will perhaps be raised that men can feign. Thus, though a criminal at the bar may show physical symptoms of mental disorder, one would nevertheless be justified in wondering whether his mental confusion were "real" or only simulated. One must note that in the case of the simulator, only some of the conditions are fulfilled which verify the statement "This man is mentally unbalanced" – those, namely, which are most accessible to direct observation. A more penetrating examination – which should in principle take into account events occurring in the central nervous system – would give a decisive answer; and this answer would in turn clearly rest on a physicalistic basis. If, at this point, one wished to push the objection to the point of admitting that a man could show *all the "symptoms"* of a mental disease without being "really" ill, we reply that it would be absurd to characterize such a man as "really normal"; for it is obvious that by the very nature of the hypothesis we should possess no criterion in terms of which we could distinguish this man from another who, while exhibiting the same bodily behavior down to the last detail, would "in addition" be "really ill." (To put the point more precisely, one can say that this hypothesis contains a *logical contradiction*, since it amounts to saying, "It is possible that a statement should be false even when the necessary and sufficient conditions of its truth are fulfilled.")

Once again we see clearly that the meaning of a psychological proposition consists merely of the function of abbreviating the description of certain modes of physical response characteristic of the bodies of man and the animals. An analogy suggested by O. Neurath may be of further assistance in clarifying the logical function of psychological statements.[8] The complicated statements that would describe the movements of the hands of a watch in relation to one another – and relative to the stars – are ordinarily summed up in an assertion of the following form:

---

[8] "Soziologie im Physicalismus," *Erkenntnis*, 2 (1931), 393, particularly p. 411.

"This watch runs well (runs badly, etc.)." The term "runs" is introduced here as an auxiliary defined expression which makes it possible to formulate briefly a relatively complicated system of statements. It would thus be absurd to say, for example, that the movement of the hands is only a "physical symptom" which reveals the presence of a running which is intrinsically incapable of being grasped by physical means, or to ask, if the watch should stop, what has become of the running of the watch.

It is in exactly the same way that abbreviating symbols are introduced into the language of physics, the concept of temperature discussed above being an example. The system of physical test sentences *exhausts* the meaning of the statement concerning the temperature at a place, and one should not say that these sentences merely have to do with "symptoms" of the existence of a certain temperature.

Our argument has shown that it is necessary to attribute to the characteristic concepts of psychology the same logical function performed by the concepts of "running" and of "temperature." They do nothing more than make possible the succinct formulation of propositions concerning the states or processes of animal or human bodies.

The introduction of new psychological concepts can contribute greatly to the progress of scientific knowledge. But such an introduction is accompanied by a danger – that, namely, of making an excessive and, consequently, harmful use of new concepts, which may result in questions and answers devoid of sense. This is frequently the case in metaphysics, notably with respect to the notions which we formulated in section II. Terms which are abbreviating symbols are taken to designate a special class of "psychological objects," and thus one is led to ask questions about the "essence" of these objects and of how they differ from "physical objects." The time-worn problem concerning the relation between mental and physical events is also based on this confusion concerning the logical function of psychological concepts. Our argument, therefore, enables us to see that *the psychophysical problem is a pseudoproblem*, the formulation of which is based on an inadmissible use of scientific concepts; it is of the same logical nature as the question, suggested by the example above, concerning the relation of the running of the watch to the movement of the hands.[9]

VI

In order to bring out the exact status of the fundamental idea of the physicalistic interpretation of psychology (or logical behaviorism), we shall contrast it with certain theses of psychological behaviorism and of classical materialism, which appear to be closely related.[10]

1. Logical behaviorism claims neither that minds, feelings, inferiority complexes, voluntary actions, etc., do not exist, nor that their existence is in the least doubtful. It insists that the very question as to whether these psychological constructs really exist is already a (pseudoproblem,) since these notions in their "legitimate use" appear only as abbreviations of physicalistic statements. Above all, one should not interpret the position sketched in this paper as amounting to the view that we can know only the "physical side" of psychological processes and that the question as to whether there are mental phenomena behind the physical processes falls beyond the scope of science and must be left either to faith or to the conviction of each individual. On the contrary, the logical analyses originating in the Vienna Circle, of which one of the consequences is the physicalistic analysis of psychology, teach us that every meaningful question is, in principle, capable of a scientific answer. Furthermore, these analyses show that that which is, in the case of the mind-body problem, considered as an object of belief is absolutely incapable of being expressed by a factual proposition. In other words, there can be no question here of an "article of faith." Nothing can be an object of faith which cannot, in principle, be known.

[9] Carnap, *Der logische Aufbau der Welt*, pp. 231–6; Carnap, *Scheinprobleme in der Philosophie*. See also note 4 above.

[10] A careful discussion of the ideas of so-called internal behaviorism is to be found in W. Koehler, *Psychologische Probleme* (Berlin: Springer, 1933). See particularly the first two chapters [translated under the title *Gestalt Psychology*].

2. The thesis developed here, though related in certain ways to the fundamental idea of behaviorism, does not demand, as does the latter, that psychological research restrict itself methodologically to the study of the responses made by organisms to certain stimuli. It by no means offers a theory belonging to the domain of psychology, but rather it offers a logical theory about the propositions of scientific psychology. Its position is that the latter are without exception physicalistic statements – by whatever means they may have been obtained. Consequently, it seeks to show that if in psychology only physicalistic statements are made, this is not a limitation because it is logically *impossible* to do otherwise.

3. In order for logical behaviorism to be acceptable, it is not necessary that we be able to describe the physical state of a human body which is referred to by a certain psychological statement – for example, one dealing with someone's feeling of pain – down to the most minute details of the phenomena of the central nervous system. No more does it presuppose a knowledge of all the physical laws governing human or animal bodily processes; nor a fortiori is the existence of rigorously deterministic laws relating to these processes a necessary condition of the truth of the behavioristic thesis. At no point does the above argument rest on such a concrete presupposition.

VII

In concluding, I should like to indicate briefly the clarification brought to the problem of the division of the sciences into totally different areas, by the method of the logical analysis of scientific statements that are applied above to the special case of the place of psychology among the sciences. The considerations we have advanced can be extended to the domain of sociology, taken in the broad sense as the science of historical, cultural, and economic processes. In this way one arrives at the result that every sociological assertion which is meaningful – that is to say, in principle verifiable – "has as its subject-matter nothing else than the states, processes and behavior of groups or of individuals (human or animal), and their responses to one another and to their environment,"[11] and consequently that every socio-logical statement is a physicalistic statement. This view is characterized by Neurath as the thesis of "social behaviorism," which he adds to that of "individual behaviorism" which we have expounded above. Furthermore, we can show that every proposition of what are called the "sciences of mind and culture" is a sociological proposition in the above sense, provided it has genuine content. Thus we arrived at the "thesis of the unity of science":

The division of science into different areas rests exclusively on differences in research procedures and direction of interest; *one must not regard it as a matter of principle. On the contrary, all the branches of science are in principle of one and the same nature; they are branches of the unitary science, physics.*

PHYSICALISM

VIII

The method of logical analysis which we have attempted to explicate in clarifying, by way of example, the propositions of psychology leads, as we have been able to show only too briefly for the sciences of mind and culture, to a "physicalism" based on logic (Neurath): *Every proposition of the above-mentioned disciplines – and, in general, of experimental science as a whole – which is not merely a meaningless sequence of words is translatable, without change of content, into a proposition in which appear only physicalistic terms – that is, consequently, a physicalistic proposition.*

This thesis frequently encounters a strong opposition arising from the idea that such analyses violently and considerably reduce the richness of the life of mind or spirit, as though the aim of the discussion were purely and simply to eliminate vast and important areas of experience. Such a conception comes from a false interpretation of physicalism, the main elements of which we have already examined in section VII above. As a matter of fact, nothing can be more remote from a philosophy which has

---

[11] Carnap, "Die physikalische Sprache," *Erkenntnis* 2 (1931), 451. See also O. Neurath, *Empirische Soziologie* (Vienna: Springer, 1931), the fourth monograph in the series *Schriften zur wissenschaftlichen Weltauffassung.*

the methodological attitude we have characterized than the making of decisions, on its own authority, concerning either the truth or falsity of particular scientific statements or the desire to eliminate any matters of fact whatsoever. *The subject-matter of this philosophy is limited to the form of scientific statements and the deductive relationships obtaining between them.* It is led by its analyses to the thesis of physicalism, and it establishes on purely logical grounds that a certain class of venerable philosophical "problems" consists of pseudoproblems. It is certainly to the advantage of the progress of scientific knowledge that these imitation jewels in the coffer of scientific problems be known for what they are and that the intellectual powers which have until now been devoted to a class of senseless questions which are by their very nature insoluble become available for the formulation and study of new and fruitful problems. That the method of logical analysis stimulates research along these lines is shown by the numerous publications of the Vienna Circle and those who sympathize with its general point of view (H. Reichenbach, W. Dubislav, and others).

In the attitude of those who are so bitterly opposed to physicalism, an essential role is played by certain psychological factors relating to individuals and groups. Thus the contrast between the concepts (*Gebilde*) developed by the psychologist and those developed by the physicist, or, again, the question as to the nature of the specific subject-matter of psychology and the cultural sciences (which present the appearance of a search for the essence and unique laws of "objective mind"), is usually accompanied by a strong emotional coloring which has come into being during the long historical development of the "philosophical conception of the world," which was considerably less scientific than normative and intuitive. These emotional factors are still deeply rooted in the picture by which our epoch represents the world to itself. They are protected by certain affective dispositions which surround them like a rampart, and for all these reasons they appear to us to have a verifiable content which a more penetrating analysis shows to be impossible.

A psychological and sociological study of the causes of the appearance of these "concomitant factors" of the metaphysical type would take us beyond the limits of this study;[12] but without tracing the appearance of these "concomitant factors" back to its origins, it is possible to say that if the logical analyses sketched above are correct, the fact that they necessitate at least a partial break with traditional philosophical ideas which are deeply dyed with emotion can certainly not justify an opposition to physicalism – at least if one admits that philosophy is to be something more than the expression of an individual vision of the world, that it aims at being a science.

---

[12] O. Neurath has made interesting contributions along these lines in *Empirische Soziologie* and in *"Soziologie im Physikalismus"* (see above note 8), as has R. Carnap in his article "Überwindung der Metaphysik durch logische Analyse der Sprache," *Erkenntnis*, 2 (1931), 219–241.

Chapter 10
Schlick and Neurath
Foundation vs. Coherence in Scientific Knowledge[1]

(1983)

[Quote H. --
"Why didn't
anybody
tell me
about
neurath?"
(gm?)]

1. INTRODUCTION

This symposium is dedicated to the ideas of Moritz Schlick and Otto Neurath. I am grateful for being able to count them among those thinkers who had an essential influence on my own philosophical development.

In the fall of 1929, I participated with great interest in Schlick's seminar and in his lecture on the philosophy of nature, and I heard the debates between him and Neurath in the lively meetings of the Vienna Circle. I visited lectures given by Neurath in Vienna and later in Berlin and Chicago, and I had many stimulating, sometimes exciting, discussions with him. I think of these very different thinkers, different in temperament and philosophical style, with deep respect and gratitude.

In my contribution to this symposium, I will not evoke personal memories; rather, I will investigate the opposing views that Schlick and Neurath held on one of the major problems of epistemology.

The problem I have in mind is the question under which conditions – or given which reasons – should we accept empirical statements, especially the statements of the empirical sciences. I will call Schlick's and Neurath's views foundational and coherentist, for short.

[1] This paper is based on studies that were supported by the National Science Foundation, Research Grant No. SES-8025399.

181

Both Schlick and Neurath were empiricists. In the brochure "The Scientific Conception of the World: The Vienna Circle," which was published in honor of Moritz Schlick by Carnap, Hahn, and Neurath in 1929,[2] the basic conviction of the group is said to be *"empiricist and positivist*: there is knowledge only from experience, which rests on what is immediately given. This sets the limit for the content of legitimate science."[3]

Obviously, this statement is vague and programmatic. It was one of the main aims of the Vienna Circle to render it more precise and to justify it. Schlick's and Neurath's ideas on this subject give us two very different interpretations of the basic empiricist principle. Schlick developed his ideas in his essay "On the Foundation of Knowledge,"[4] in which he also critically discussed ideas that Neurath had proposed in some of his work. Neurath responded in his essay "Radical Physicalism and the 'Real World.'"[5] In a small article ("On the Logical Positivists' Theory of Truth") that was published soon thereafter, I tried to defend Neurath's ideas by clearing up some misunderstandings.[6]

---

[2] *Wissenschaftliche Weltauffassung: Der Wiener Kreis.* This brochure originally appeared in a series called "Publications of the Ernst Mach Society" (Vienna: Artur Wolf, 1929). The text, by Otto Neurath, is reprinted in Rudolf Haller and Heiner Rutte, eds., *Gesammelte philosophische und methodologische Schriften* (Vienna: Hölder-Pichler-Tempski, 1981), pp. 299–336. For an English translation see Marie Neurath and Robert S. Cohen, eds., *Otto Neurath, Empiricism and Sociology* (Dordrecht: Reidel, 1973), pp. 299–318.

[3] Neurath and Cohen, eds., p. 309.

[4] Schlick's paper, "Über das Fundament der Erkenntnis," was first published in *Erkenntnis* 4 (1934), 79–99. It has been translated by Peter Heath as "On the Foundation of Knowledge," and it is reprinted in Henk L. Mulder and Barbara F. B. van de Velde-Schlick, eds., *Moritz Schlick: Philosophical Papers*, vol. 2 (Dordrecht: Reidel, 1979), pp. 370–88. All references are to Mulder and van de Velde-Schlick, op. cit.

[5] An English translation of Neurath's paper, "Radikaler Physikalismus und 'Wirkliche Welt,'" originally published in *Erkenntnis* 4 (1934), 346–62, can be found in Otto Neurath, "Radical Physicalism and the 'Real World,'" in *Philosophical Papers 1913–1946*, which was edited and translated by Robert S. Cohen and Marie Neurath (Dordrecht: Reidel, 1983), pp. 100–14. All references are to Cohen and Neurath, op. cit.

[6] *Analysis* 2 (1935), 49–59. (Chapter 1 here.)

Schlick responded in his article "Facts and Propositions."[7] Yet the disagreement between these two basic approaches has not been clearly resolved.

"TRUTH"

2. SCHLICK'S FOUNDATIONAL VIEW

Schlick thought that the purpose of science is "to provide a *true* account of the facts."[8] Thus, Schlick aims at a version of empiricism according to which a criterion of truth for scientific sentences can be formulated by referring to the data of immediate experience.

In order to achieve such a version of empiricism, Schlick relies on the hypothetical-deductive model of testing scientific theories. According to it, we first deduce certain consequences from the hypothesis that we want to test. These deduced sentences *H~D* should describe immediately observable kinds of events, especially directly observable experimental results like the position of the indicator on a measurement instrument or the change in color in some chemical reaction. Then we examine whether the observation sentences or predictions we derived from the hypothesis are indeed in agreement with the results of appropriate experiments. Such results are directly observed and recorded in observation sentences. I will call these sentences directly advanced observation sentences in order to distinguish them from those sentences that report our prediction of some observation and were hypothetically deduced from some hypothesis.

If these directly advanced observation sentences are in conflict with the observational predictions that we derived from the hypothesis, then the hypothesis is "falsified" by the empirical findings. If the predictions of what we will observe are in accordance with what we actually observe, then the hypothesis is, though not verified, partially confirmed. So long as testing only provides us with confirming results, we believe, according to

---

[7] *Analysis* 2 (1935), 65–70; reprinted in Mulder and van de Velde-Schlick, op. cit., pp. 400–4.
[8] Schlick, "On the Foundation of Knowledge," p. 374.

Schlick, that the hypothesis is one that we have "guessed correctly."[9]

It might seem as if this idea had led to an (at least partial) empiricist criterion of truth for scientific hypotheses – namely, to the following necessary but not sufficient condition: A hypothesis is true only if it is not in conflict with any directly advanced observation sentence.

But this obviously presupposes that each directly advanced observation sentence is true. If that were not the case, a hypothesis contradicting such a sentence need not at all be false. If all directly advanced observation sentences were really true, then they would indeed form a firm and unshakable foundation of knowledge. In particular, each hypothesis that would contradict these sentences would be falsified in the semantically strict sense: The conflict between hypothesis and observation sentence would prove the hypothesis to be *false*.

If, on the other hand, a directly advanced observation sentence might itself be false, then it falsifies the hypothesis in conflict with it only in a relative sense – namely, in the sense that if the observation sentence is true, then the hypothesis has to be false.

Schlick is looking for a criterion of truth or falsity for scientific hypotheses. So the question suggests itself whether the testing of scientific hypotheses can be founded on a certain class of observation sentences that, once advanced, were immediately certain – and, as Schlick puts it, "cannot be shaken" and "remain unaltered" and are such that with them all other scientific sentences "must be brought into agreement."[10]

Asking this question, Schlick is completely aware of the fact that the sentences with which scientists describe their observations are always open to the possibility of error and correction. He adds that the same holds for protocol sentences (i.e., for directly advanced observation sentences) and that the possibility of error can never be excluded, not even in the case of sentences advanced by ourselves.[11]

But Schlick goes on to say that when we order sentences

---

[9] Schlick, "On the Foundation of Knowledge," p. 381.
[10] Ibid., p. 377.   [11] Ibid., p. 373.

according to their certainty, a subclass of the sentences that I advance myself will be at the top of this order. These sentences – which are "immune from all doubt" – "give expression to a matter of personal 'perception' or 'experience' . . . that lies *in the present*."[12] Schlick calls such sentences 'affirmations' (*Konstatierungen*). They are always of the form "Here now so-and-so" (e.g., "Here now two black spots coincide" or "Here now blue is bounded by yellow"). "Affirmations," according to Schlick, can only be expressed in connection with a referring gesture.[13] Thus, in Schlick's view, a real "affirmation" cannot be noted down. If you write down the words "here" and "now," they lose their meaning; if you replace them by a specific description of place and time, the whole statement loses the certainty characteristic for "affirmations." This certainty is, according to Schlick, due to the fact that when one verifies an "affirmation," understanding its meaning already amounts to acknowledging its truth. (In his words: "Along with their meaning I simultaneously grasp their truth."[14])

Schlick emphasizes that "affirmations" are not identical with what "could properly be called 'protocol sentences.'" The latter are sentences that could be written down, and they share the uncertainty of all scientific hypotheses. "Affirmations" are not potocol sentences; they are rather "the occasion for framing them."[15] In this sense can they be regarded as "the ultimate origin of all knowledge."[16] On the other hand, "affirmations" are, in Schlick's view, also the endpoint of the scientific testing of all hypotheses. This is because if you derive some predictions about what you will observe from a hypothesis, the question emerges whether these predictions will really come true. "In every single case of verification or falsification, an 'affirmation' answers unambiguously with *yes* or *no*, with joy of fulfillment or disillusion. The affirmations are final . . . They are an absolute end, and in them the current task of knowledge is fulfilled."[17] Schlick thinks that "the problem of the 'foundation' will automatically transform itself into that of the unshakeable points of contact

---

[12] Ibid., p. 379.   [13] Ibid., p. 385.   [14] Ibid., p. 385.
[15] Ibid., p. 381.   [16] Ibid., p. 381.   [17] Ibid., p. 383.

Schlick's inconsistency

between knowledge and reality."[18] These unshakable points are Schlick's "affirmations."

To me it seems clear that Schlick is faltering between two incompatible views of "affirmations." On the one hand, Schlick talks about "affirmations" as observation sentences. He tries to justify his view that they are true and certain. On the other hand, Schlick denies "affirmations" the status of sentences that could be written down or – as we might say expressing his view – of sentences that can be somehow put into words. But then it does not make sense to call them true or certain. Schlick, I think, hints at this second view when he characterizes "affirmations" as the occasion for framing protocol sentences. According to this view, "affirmations" are psychological events that consist of the fact that an observer perceives or seems to perceive some event such as the coincidence of a point of light with some black line in his visual field. Obviously, such an event is neither true nor false; it is neither certain nor uncertain: Either it occurs or it does not. But it can play an important *causal* role in the process that leads an observer to use a specific protocol sentence in his report about his experimental findings. I will come back later to this causal aspect of the notion of "affirmation."

Schlick calls his view a form of correspondence theory of truth, according to which, schematically spoken, the truth of a sentence would consist in its agreement with the facts. He puts his view in opposition to Neurath's ideas, which he calls a coherence theory, according to which the truth of a sentence consists exclusively in the agreement of this sentence with the system of other sentences.[19]

3. NEURATH'S HOLISTIC ACCOUNT OF SCIENTIFIC KNOWLEDGE

In a series of essays Neurath had indeed proposed the view – of which he gave a renewed statement in his answer to Schlick –

[18] Schlick, "On the Foundation of Knowledge," p. 387.
[19] Ibid., pp. 374–7.

[Handwritten annotations in top margin: "⇒ [THIS is MUCH CLOSER TO KUHN's views] — AN EXAMPLE OF K.'s DEBT TO THE "LEFT WING" CPS.]"]

that to test a scientific hypothesis is to compare it with other sentences, especially with directly advanced observation sentences, which he called protocol sentences. Neurath thought that these protocol sentences have a particular and rather complicated form, which need not concern us here.

Neurath emphasized again and again that to test a scientific hypothesis is not to compare it with the "facts" or with the "real world," but to compare it with other sentences. For example, Neurath says,

> What is under discussion is science as a system of statements. *Statements are compared with statements*, not with "experiences," not with "the world," nor with anything else. All these meaningless *duplications* belong to a more or less refined metaphysics and are therefore to be rejected. Each new statement will be confronted with the totality of existing statements that have already been harmonized with each other. *A statement is called correct, if it can be incorporated* in this totality.[20]

Neurath's rejection of the idea that hypotheses can be compared with "reality" arose from his antimetaphysical attitude; he thought that this idea would only lead to subjectless pseudoproblems. On the other hand, though, he remained an empiricist. Neurath interpreted the basic empiricist idea as follows: "Thus for us striving after knowledge of reality is reduced to striving to establish agreement between the statements of science and as many protocol sentences as possible. But this is *very much*; in this rests *empiricism*."[21] In the same spirit, Neurath writes, "The

---

[20] Otto Neurath, "Soziologie im Physikalismus," *Erkenntnis* 2 (1931), 393–431. The same view is expressed in Neurath's "Radical Physicalism and the 'Real World,'" op. cit.; see especially section 3, "Conformity with Reality," pp. 107–10. The first mentioned paper has been translated as "Sociology in the Framework of Physicalism" and can also be found in Otto Neurath, *Philosophical Papers 1913–1946*, edited and translated by Robert S. Cohen and Marie Neurath (Dordrecht: Reidel Publishing Company, 1983), pp. 58–90. The quotation is from p. 66.

[21] Neurath, "Radical Physicalism and the 'Real World,'" p. 109.

renunciation of confronting all predictions in the last resort with protocol statements can probably not be maintained without endangering the fundamental *empirical* standpoint."[22]

Neurath emphasizes that, despite their essential role in testing empirical statements, protocol sentences do not form a firm and unshakeable foundation of knowledge. For example, in his reply to Schlick, Neurath says, "All real statements of science, including those protocol sentences that are used for verification, are selected on the basis of decisions and can be altered in principle."[23] Thus, all accepted scientific statements – protocol sentences as well as "nonprotocol sentences" – are, as Neurath emphasizes, open to the possibility of revision and rejection. Let us look in more detail at some of the reasons for this claim.

In respect to the revisability of nonprotocol sentences Neurath refers, for examples, to the ideas of Duhem, who, already at the beginning of the century, clearly showed that a single hypothesis, viewed in isolation of others, is not empirically testable at all. Take, for example, this hypothesis: "All electrons bear the same electric charge." In order to derive any predictions about directly observable experimental findings from this hypothesis, we have to use a comprehensive system of additional hypotheses. These additional hypotheses figure as further premisses in such a derivation. Consequently, if a so-derived prediction about some observation contradicts a directly advanced observation statement, then the latter falsifies the set of used hypotheses, but it does not falsify the single hypothesis we wanted to test. Logical consistency can be regained either by rejecting the hypothesis we wanted to test or by rejecting one or another of the other hypotheses which we used as further premisses. There is no *experimentum crucis* for a single hypothesis, no possibility of even a relative falsification by observation statements. A system of hypotheses can always be adjusted to unfavorable experimental findings in many different ways. In fact, the history of science provides many examples in which negative experimental findings led not to a rejection of the tested hypothesis but to a rejection of

---

[22] Neurath, "Radical Physicalism and the 'Real World,'" p. 107f.
[23] Ibid., p. 102.

auxiliary hypotheses which served as essential premisses in the derivation of predictions.

As already mentioned, Neurath expanded this idea further by claiming that a contradiction between a theory and directly advanced protocol sentences can also be resolved by rejecting the latter.[24] Haller has called this important idea the Neurath-principle.[25] Popper and, soon, also Carnap took a very similar stance on this matter.

This way of rejecting protocol sentences does, in fact, occur in science. It is not a rare phenomenon that observational findings that contradict a hitherto strongly and manifoldly confirmed theory are rejected. In these cases the principle by which a narrow empiricism could be characterized does not hold – namely, the principle that all scientific theories have to be adjusted to fit the data of direct observation. It is not always the case that these data decide the acceptability of a theory; not uncommonly, theories decide whether the given data are acceptable.

In the early Vienna Circle, one tried to establish the empirical meaningfulness or the empirical content of singular hypotheses by reference to observation sentences that are deducible from them. We could call this an atomic view of empirical knowledge. According to it, empirical knowledge is constructed by single statements, each of which has its own empirical content. The view which Neurath advanced with great emphasis – even if he did not always formulate or justify it with the utmost precision – and which by now has pretty much succeeded could be called a holistic view. We could characterize it by the following slogans: To an individual scientific hypothesis, we can assign neither unequivocal testing methods nor a definite empirical content. The exact content of a hypothesis and how it can be tested depends on the whole theoretical system in which the hypothesis and the concepts it contains are systematically used, and – if we take this dependence strictly – it also depends on the totality

---

[24] Compare the explicit formulation of this idea in Neurath, "Protokollsätze," *Erkenntnis* 3 (1932), 204–14; this article was translated in Cohen and Neurath, eds., "Protocol Statements," *Empiricism and Sociology*, pp. 91–9; see p. 95.

[25] Rudolf Haller, "New Light on the Vienna Circle," *The Monist* 65 (1982), 25–35; quotation from p. 33.

189

of other theories that are at the time accepted in science. The latter holds because such further theories are typically involved when a hypothesis is tested by some experiment. For example, optical theories are used to test statements of Newton's celestial mechanics.

Thus, the question of scientific acceptability or credibility refers not to singular statements but to whole systems of sentences, systems which contain protocol sentences as well as nonprotocol sentences. And the requirement that the latter should be supported by the former must be replaced by a more general condition that demands that all sentences of a system stand in an appropriate sort of connection.

Schlick calls this connection coherence, for short, and remarks that, in Neurath's view, coherence can plausibly only amount to logical consistency. He goes on to reject this coherence view with the remark that, according to it, a book of consistent fairy tales and a book of physics would be on a par – both would have to be accepted as systems of true scientific sentences.[26]

In fact, for Neurath logical consistency is not the only requirement for a system of sentences to be scientifically acceptable. Such a system would also have to fulfill his empiricist requirement – namely, that nonprotocol sentences have to be in agreement with as many protocol sentences as possible. And Neurath adds that also the "simplicity" of the ways in which sentences can be connected is one of the factors that partly determine the choice of a scientific system of sentences.[27]

But all this admits, in principle, of a large variety of acceptable sets of sentences – that is, of acceptable scientific world pictures – each one of which can essentially differ from others in its protocol sentences as well as in its nonprotocol sentences. The scientists have to choose among them. Thus, Neurath strictly opposes the idea that there is one true picture of the world.

Let me also add, briefly, that Schlick's view cannot avoid a pluralism of admissible theories. This is because each set of observational data, however large it might be, can be subsumed under

---

[26] Schlick, "On the Foundation of Knowledge," p. 376.
[27] Otto Neurath, "Radical Physicalism and the 'Real World,'" p. 105.

different theories, some of them even incompatible ones. The data we gather from experience never determine, for purely logical reasons, a single theory, let alone "the true" theory.

4. TRUTH VS. CONFIRMATION

As mentioned before, Neurath refuses to characterize the truth of sentences in terms of their agreement with facts. Instead, he proposes a pragmatic-sociological definition of the words "true" and "false": *True* we call the set of sentences we accept and every sentence that is in agreement with the consequences of the accepted set or could be incorporated in it; we call a sentence *false* if it contradicts the accepted set.[28]

But this proposal confuses two very different sorts of notions, both of which are of essential importance for the logic and methodology of science. On the one hand we have semantic notions like truth and falsity, and on the other hand we have epistemological notions like scientific credibility or the acceptability of empirical statements.

Other empiricist-oriented thinkers, including myself, shared this confusion. At that time the notion of truth was regarded with suspicion in the circles of logical positivists. It seemed either superfluous or metaphysical.

But these doubts evaporated when Tarski developed his precise semantic theory of truth, which was enthusiastically advocated by Carnap. With the help of Tarski's theory, Carnap could completely resolve the conceptual confusion mentioned above.[29]

[28] Otto Neurath, "Einzelwissenschaften, Einheitswissenschaft, Pseudorationalismus" (1936), translated as "Individual Sciences, Unified Science, Pseudo-Rationalism," in Cohen and Neurath, eds., *Empiricism and Sociology*, pp. 132–8; see p. 135.

[29] Alfred Tarski, "Der Wahrheitsbegriff in den formalisierten Sprachen" (orig. 1933), *Studia Philosophica* 1 (1936), 261–405; translated as "The Concept of Truth in Formalized Languages," in Alfred Tarski, *Logic, Semantics, Metamathematics* (Oxford: Clarendon Press, 1956), pp. 152–278. In respect to the conceptual confusion, see Rudolf Carnap, "Truth and Confirmation," in H. Feigl and W. Sellars, eds., *Readings in Philosophical Analysis* (New York: Appleton-Century-Crofts, 1949), pp. 119–27.

Some of the main differences between these two groups of notions are the following: (i) A sentence is timelessly true or false, independent of whether it will ever be tested or not. Whether a sentence is confirmed or acceptable, on the other hand, depends on the test results available at the time, and confirmation can thus change over time. Confirmation and acceptability are relative epistemological notions; they hold of a sentence relative to some body of information. (ii) It is possible that a true sentence is only weakly supported by the available information and is, relative to this information, not acceptable, while false sentences could be well supported. (iii) It is true for any sentence that either this sentence or its negation is true, but it does not hold that either this sentence or its negation is well-supported.

The semantic notion of truth is indispensable in logic, and it cannot be replaced by the epistemological notion of acceptability. The latter notion is of great interest within the philosophy of science, but it is itself very much in need of further clarification.

Neurath's holistic view of knowledge and his idea to give a sociological description of "true" hypotheses should thus not be regarded as a theory of truth. Rather, they constitute a pragmatic interpretation of the acceptance and the acceptability of scientific systems of sentences.

For some time, the questions Neurath dealt with have been – especially under the influence of Thomas Kuhn's ideas – intensely discussed as the problem of how to choose a scientific theory. Roughly put, the question is what factors decide which of two competing theories should be preferred and should be accepted for the purposes of future research.

Kuhn deals thoroughly with this question in his treatment of scientific revolutions. He says that scientists prefer those theories that fulfill certain conditions – conditions that are regarded as important – better than their competitors. These conditions – I will call them desiderata – are the characteristics of a good theory. Typical examples of such desiderata are the following: agreement with the results of experiments (this is, again, the basic empiricist requirement), applicability to a wide range of diverse phenomena, the capacity to make correct predictions in respect to new kinds of phenomena, and simplicity. Kuhn thinks that none of these desiderata – not to speak of their totality – can be formulated as a precise criterion that unequivocally favors one of the competing theories.[30] He thinks of these desiderata as values. During their education and their professional career, scientists have made these values their own. Though scientists share these values, they do not understand or apply them in the very same way. Differences in opinion in regard to the merits of competing theories can occur among the most competent scientists without any of them having violated some clear methodological standards.

Neurath rejected the idea of a system of precise methodological rules that, for example, tells us which hypotheses to accept. He called such an idea "pseudorationalism" and proposed instead a behavioristic-sociological description of the scientific enterprise. The danger of pseudorationalism, Neurath says, occurs, for example, when one thinks that one could replace the decisions made in the scientific practice by a "calculus of the logic of science." In fact, this is possible when we deal with logical or mathematical deductions, "where we can regard scholars as a sort of automata that detect contradictions and deduce consequences."[31] But in the case of choosing between two competing empirical theories, Neurath continues, two scientists can rely on different considerations that will not lead to clear agreement. A more detailed inquiry into these questions belongs to the subject-matter of the "behavioral science of scientists," and the danger of pseudorationalism lies in the temptation to subsume the considerations of scientists under the precise rules of a logic of science "without having the necessary data as a foundation."[32]

Neurath rejects all attempts to characterize the procedures of empirical science by precise and general rules, be these the rules

---

[30] J. von Kries shared a similar skepticism in regard to whether we can attribute precise probabilities to theories; see his book *Die Prinzipien der Wahrscheinlichkeitsrechnung* (Freiburg: J. C. Mohr, 1886; Türbinger: J. C. Mohr, 1927) pp. 26, 29f.; and Carnap's thorough discussion of the problem and of Kries's view in his *Logical Foundations of Probability* (Chicago: University of Chicago Press, 1950), pp. 219–33.

[31] Neurath, "Individual Sciences," p. 136f.   [32] Ibid.

of verification, of falsification, of induction, or of theory testing. But he explicitly allowed for the possibility that precise rules may be valuable in the narrow domain of specific research problems.[33]

Neurath's rejection of "pseudorationalism" is related to an idea advanced by Duhem already thirty years before Neurath. Duhem emphasized that when a theory conflicts with observational findings, there are no precise rules that determine in which way the theoretical system has to be altered. Such a decision must be left to common sense.[34]

In their pragmatic-sociological orientation, Neurath's ideas bear a distinct affinity to those of Kuhn. Neurath expresses and argues for his views often in a programmatic and sketchy style that does not fully satisfy the standards for a systematic and well-argued exposition. Kuhn, on the other hand, and other contemporary philosophers of science who share a similar attitude try to give a detailed account of the main factors that determine the choice of scientific theories, supporting their case with relevant facts from the history of science.

5. METHODOLOGY OF EMPIRICAL SCIENCE: DESCRIPTIVE OR NORMATIVE?

The considerations above lead to the question of whether a pragmatic-sociological–oriented study of science is limited to an empirical-descriptive theory of how scientific research is done ("the behavioral science of scientists") or whether it could also establish norms for correct and rational scientific procedures. Given such a theory, is it possible to criticize the behavior of scientists – for example, if they just make up the results of their experiments – as violations of norms that constitute the scientific method?

[33] Otto Neurath, "Pseudorationalismus der Falsifikation," *Erkenntnis* 5 (1935), 353–65, translated as "Pseudorationalism of Falsification," in Cohen and Neurath, eds., *Empiricism and Sociology*, pp. 121–31.

[34] Pierre Duhem, *La théorie physique: Son objet et sa structure* (Paris: Rivière, 1906, 1914). This has been translated as *The Aim and Structure of Physical Theory* (Princeton: Princeton University Press, 1954). See Part 2, Chapter 6, sec. 10.

It might seem as if scientists, in Neurath's view, could not make any mistake, because Neurath defines "true" sentences as those which are accepted by scientists. But Neurath obviously did not want to give his ideas such a narrow reading. Remember that he emphasized that each statement accepted in science at some point could also later be rejected, which would allow us then to call this statement false. Before I deal with the role of norms in Neurath's picture of science, I want to point out that other empiricists of those days vehemently rejected a purely descriptive or "naturalistic" view of the methodology of science. Popper's criticisms of such a view were especially explicit and insistent.

Popper called the methodological principles of his own *Logic of Scientific Discovery* "conventions" – conventions that could be regarded as the rules of the game of science. But purely conventional rules could not claim to be a methodology of empirical science if they did not do justice in an appropriate way to the characteristic features of what is commonly regarded as scientific research – as, I think, Popper would agree.[35]

A theory of scientific knowledge has to contain descriptive as well as normative aspects; it has to be founded on empirical information and normative standards. The philosophical development or explication of methodological principles requires a mutual adaptation of both kinds of considerations.

This can be illustrated by the step-by-step transformation of many ideas that were dealt with in the Vienna Circle. For example, the norm that any empirically contentful hypothesis must be strictly verifiable by observation statements was abandoned early on. The main reason for this revision was the following empirical fact: In empirical science more or less unrestricted universal sentences are accepted – for example, alleged laws of nature. But these universal sentences cannot, for logical

[margin note: VC ABANDONS VERIF. PRINCIPLE]

---

[35] See, for example, Popper's remarks on the justification of his methodological conventions in sections 9, 10, and 11 of his book *Logik der Forschung* (Vienna: Julius Springer, 1935). This book was translated by Popper as *The Logic of Scientific Discovery* (London: Hutchinson, 1958). Compare also his more explicit remarks on pp. 1030–37 in Paul R. Schilpp, ed., *The Philosophy of Karl Popper* (La Salle, IL: Open Court, 1974).

reasons, be strictly verified. On the other hand, analytically oriented empiricists denied to neovitalism the status of a scientific
theory – because it did not fulfill an even liberal demand of
empirical testability – despite the fact that respected biologists
accepted this doctrine.

In the first case a norm was abandoned because of how scientists work; in the second case a system of ideas developed by scientists was rejected as unscientific because it did not fit a retained
norm.

The principles of a methodology of the empirical sciences
are neither exclusively descriptive statements (describing how
science is done) nor exclusively norms (telling us how science
should rationally be done). Furthermore, methodological principles cannot be clearly separated into purely descriptive sentences
on the one side and purely normative sentences on the other –
just as sentences in general cannot be clearly separated into analytic and synthetic ones.

Although Neurath emphasized the descriptive part of
methodology, he, without doubt, builds certain epistemological
norms into his description of empirical science. He thinks that
acceptable systems of sentences have to be, among other things,
logically consistent and deductively closed (sentences that are
deducible from accepted sentences are also part of the accepted
system of sentences). These maxims are obviously not simply
descriptions of the actual behavior of scientists: A system of
sentences could be accepted although it contains unrecognized
contradictions; likewise, a hypothesis could be withheld or even
rejected although it follows from accepted sentences in a way
unknown at the time. Neurath's basic empiricist maxim is thus
certainly not a descriptive statement about the behavioral dispositions of practicing scientists.

6. DOES NEURATH HOLD A COHERENCE THEORY?

At the end I want to come back to the question of whether
Neurath's view is, in fact, a coherence theory of truth, as Schlick
thought it to be.

I have already mentioned that Neurath's theory is certainly no theory of truth at all. It is rather a theory of scientific acceptability, although Neurath himself used the words "true" and "false" in the statement of his view.

But let us ask the following question: Is it true that Neurath accepted a coherentist view according to which the only demand on an acceptable system of hypotheses is that the sentences of this system stand in certain sorts of agreement-relations to each other?

Let me first point out that the relevant idea of what agreement or accordance or simply coherence of sentences and systems of sentences amounts to is left extraordinarily vague – we do not find any clear and general criteria of coherence. But Neurath's own statement of this idea shows that he demands more of an acceptable system of sentences than that there just be certain fitting-relations between the sentences of such a system. Remember that his basic empiricist maxim requires that the sentences of such a system be in accordance with as many protocol-sentences as possible (i.e., with sentences that scientists accept on the basis of their observations). Consequently, acceptable systems of hypotheses must contain protocol sentences. This condition by itself says nothing about what sort of *relations* have to hold between sentences. It just demands that the whole system contain sentences that are advanced under appropriate circumstances by competent observers. Neurath refers to protocol sentences in his formulation of empiricism. And these protocol sentences cannot be characterized – at least not exclusively – by their relation to other sentences. Rather, we have to understand what protocol sentences are through certain causal features of the situation in which they are advanced.

Interestingly enough, causal features also play a role in Schlick's characterization of "affirmations." As mentioned earlier, "affirmations" in Schlick's view are occasions for framing protocol sentences in the process of testing hypotheses.

Neurath's ideas are holistic, but they express just as little a pure coherence theory as do Schlick's ideas. Both systems of ideas go beyond a pure coherence view exactly because they both want to give due weight to the role experience plays in the generation of scientific knowledge. This task leads them both to empirical-causal considerations.

In order to take the empirical character of Neurath's protocol sentences into account, one has to support and to refine the idea of directly advanced protocol sentences using an empirical theory of human observation and concept formation. Studies that go in such a direction have been undertaken for some years now. Neurath, without doubt, would have heartily welcomed these more recent attempts toward a "naturalized epistemology."

Chapter 11

On the Cognitive Status and the Rationale of Scientific Methodology

(1988)

I. METHODOLOGY OF SCIENCE: DESCRIPTIVE AND PRESCRIPTIVE FACETS

1. Two Conceptions of the Methodology of Science

In the course of its long history – most strikingly in recent centuries – scientific inquiry has vastly broadened man's knowledge and deepened his understanding of the world he lives in, and the remarkable successes of predictions and technological applications based on those insights are widely acknowledged as eloquent testimony to the soundness or the "rationality" of scientific methods of research. Yet, there is no unanimity among students of the methodology of science as to whether or to what extent it is possible to formulate clear and precise rules of procedure which are characteristic of scientific research and which make science the exemplar of rationality in the pursuit of reliable knowledge. Nor is there unanimity concerning the grounds on which such rules of scientific inquiry could be established and how, accordingly, their cognitive status is to be understood: Is the methodology of science to be viewed as a descriptive study of actual scientific research or as a quasi-normative discipline aimed at formulating – in an analytic, aprioristic manner – standards of proper or rational scientific inquiry?

These questions have in recent years been the focus of an intense and fruitful debate between two schools of thought, which I will briefly refer to as the analytic-empiricist and the historic-sociological or pragmatist schools. By the former, I mean a body of ideas that, broadly speaking, has grown out of logical empiricism and the work of kindred thinkers. Among the recent protagonists of a historic-pragmatic perspective, I include such thinkers as Thomas Kuhn, Paul Feyerabend, and Larry Laudan.

In the view of analytic empiricism, it is indeed possible to formulate certain characteristic rules and standards of scientific procedure, and it is the task of the methodology of science to exhibit, by means of "logical analysis" or "rational reconstruction" or "explication," the logical structure and the rationale of scientific inquiry. The methodology of science is understood as concerned solely with certain general logical aspects of scientific research, which form the basis of its rationality; the psychological and historical facets of science as a social enterprise are excluded from its domain.

The illuminating work done in this spirit no doubt drew encouragement from a tempting analogy between the analytic methodology of science on the one hand and metamathematics on the other. The latter is concerned not with a descriptive account of mathematical research but rather with the formal characterization of correct proof and definition; it can be said to provide, in precise terms, certain objective standards for the validity of mathematical claims and procedures.

Carnap's conception of the philosophy of science as the logical analysis of the language of science and his and Popper's exclusion of psychological and sociological issues from its domain reflect a similarly analytic and aprioristic conception of the methodology of empirical science.

Methodological pragmatists, on the other hand, have urged that the formulation of standards for scientific inquiry cannot be a purely analytic matter, that such a formulation must reflect actual scientific practice rather than analytic preconceptions we may have about rational ways of pursuing knowledge. Thomas Kuhn has expressed this view, saying that

existing theories of rationality are not quite right and we must re-adjust or change them to explain why science works

200

as it does. To suppose, instead, that we possess criteria of rationality which are independent of our understanding of the essentials of the scientific process is to open the door to cloud-cuckooland. (Kuhn 1970b: 264)

Proponents of a pragmatist approach also reject the ideal entertained by some analytic empiricists who characterize scientific inquiry as governed by norms that can be expressed in precise and fully objective terms.

Turning to a closer comparison of these two conceptions of the character of methodological principles, I will now try to show that there are greater affinities between them than is suggested by the claims of their proponents and by the sketchy characterization I have just given of their views.

2. Analytic Construals of Methodological Principles

Let me preface my comparison with a familiar reminder. The "rules of scientific procedure" of which I have been speaking certainly cannot be thought of as explicit instructions that can be applied mechanically and can enable any person of normal intelligence to make scientific discoveries. As has become clear during the long evolution of the problem of induction, there are no such algorithmic rules. Scientific discovery, as Einstein was fond of emphasizing, requires creative imagination and ingenuity that cannot be replaced by the application of any mechanical rules. However, once a hypothesis or theory has been formulated, methodological rules of scientific inquiry might be expected to specify objective and unambiguous ways of appraising and possibly justifying it by reference to given observational or experimental findings. Thus, *given* a hypothesis and *given* a set of empirical data, methodological principles might determine in an unambiguous way whether the evidence does or does not support the hypothesis or what degree of rational credibility or probability the evidence confers upon the hypothesis.

The criteria provided by such methodological rules were broadly thought of as expressible in terms of purely logical relations between hypotheses and evidence sentences. This would ensure their objectivity, which might be threatened by admitting into the critical appraisal of hypotheses certain extralogical considerations of a historical or psychosociological or generally "naturalistic" kind.

Such was indeed the basic conception underlying analytic-empiricist theories of confirmation (e.g., those developed in Hempel 1943; 1945) and of logical, or inductive, probability (cf. Carnap 1950; 1963a).

A strong emphasis on explication by precise logical means is also evident in the definitions proposed by logical empiricists for many other concepts of the methodology of science; well-known examples are the characterizations of "empirically meaningful" hypotheses in terms of potential verifiability, falsifiability, or confirmability by "observation sentences" and by analytic construals of scientific explanation, the structure and function of theories, and theoretical reduction.

The search for methodological principles that do not refer to psychological or sociological considerations was prompted in part by the ideal of the objectivity of scientific procedures and claims. This would call for methodological norms which determine unambiguous answers to problems of critical appraisal so that different scientists applying them would agree in their verdicts. The methodological criteria developed by analytic empiricists – including criteria for confirmation, explanation, and the like – are couched largely in terms of purely logical concepts: This bodes well for their objectivity.

Actually, those criteria also use certain extralogical concepts, among them those of *observation sentence* and *observational term*, for the analytic-empiricist criteria characterize the testability, the rational credibility, and similar features of scientific claims in terms of certain logical relations between the claim in question and a set of data in the form of observation sentences or basic sentences. These are taken to describe phenomena whose occurrence or nonoccurrence can be established, with good intersubjective agreement, by means of direct, theory-independent observation.

Such interobserver agreement concerning observational evidence was seen as safeguarding the objectivity of science at the evidential level. As a result, the methodological norms for the appraisal of scientific claims would then be objective – that is, intersubjectively binding, since they called for precisely characterized logical relations between a hypothesis under appraisal and a body of evidence sentences that could be established with high intersubjective agreement by means of direct observation.

In recent decades, the notion of direct, theory-independent observability has rightly been subjected to severe criticism, leading to considerable modifications in the analytic-empiricist construal of the evidential basis for scientific claims. However, these developments need not detain us here, since they do not affect the question of the cognitive status of methodological principles, which is our present concern.

3. Normative and Descriptive Facets of Popper's Methodology

What is the cognitive status of methodological principles as envisaged by analytic empiricists and kindred thinkers? On what grounds are they set forth; by what means can their adequacy be appraised? Let us briefly consider the views of some representative thinkers, particularly Popper and Carnap.

Karl Popper rejects what he calls the "naturalistic" view – that is, the conception of methodology as an empirical account of the actual research behavior of scientists. Popper remarks that this conception would require a prior specification of what persons are to count as scientists, and he argues that "what is to be called a 'science' and who is to be called a 'scientist' must always remain a matter of convention or decision" (1959: 52). As for his own methodology, which properly characterizes scientific hypotheses by their falsifiability and sees scientific progress in the transition to ever more highly corroborated and ever better testable theories, Popper holds that its principles are *conventions* which "might be described as the rules of the game of empirical science" (1959: 53).

Significantly, however, Popper does not present those conventional rules as simply arbitrary: He offers arguments in their support. In particular, he argues that his methodology "is fruitful: that a great many points can be clarified and explained with its help" and that, by its consequences, "the scientist will be able to see how far it conforms to his intuitive idea of the goal of his endeavors" (1959: 55). "It is by this method, if by any," says Popper, "that methodological conventions might be justified" (1959: 55).

Clearly, then, Popper does not consider his methodological conventions arbitrary: He intends them to meet certain justificatory requirements, and these have something to do with how scientists conceive and pursue the goal of their endeavors. These justificatory grounds are highly unspecific: They are such that might also be offered in support of a proposed methodology of mathematics, for example. However, for his methodology of empirical science, Popper also adduces more specific considerations. As he tells us, his interest in methodology was stimulated by the idea that doctrines like astrology, Marxist theory of history, and both Freud's and Adler's versions of psychoanalysis were unsatisfactory attempts at theorizing: They were protected against any possible empirical refutation by vagueness of formulation, face-saving conventionalist stratagems, or being formulated so as to be totally untestable. In contrast, the general theory of relativity made far-reaching, precise, and specific predictions – and thus laid itself open to severe testing and possible falsification (1962: 33–4).

Popper regards these latter features as characteristic of genuinely scientific theories, and he, therefore, aimed at a methodology that would systematically elaborate this conception. Thus, Popper's theory has a target: to exhibit the rationale of certain kinds of theorizing that he judges to be scientific. Indeed, Popper remarks that if, in accordance with his methodology, we stipulate that science should "aim at *better and better testable* theories, then we arrive at a methodological principle. . . . whose [unconscious] adoption in the past would rationally explain a great number of events in the history of science." "At the same time," he adds, the principle "gives us a statement of the task of science, telling us that *should* in science be regarded as *progress*" (1979: 356).

Thus, while Popper attributes to his methodological principles a normative character, he assigns to them, in effect, an empirical-explanatory role as well. This role appears in the justificatory reasons adduced for them. Among these are the empirical claims that theories generally viewed as scientific – such as Newton's or Einstein's – are qualified as scientific by his standards; that certain others, which are antecedently considered as nonscientific, are ruled out by his criteria; and, very significantly, that many actual events in the history of science could be explained by the assumption that scientists in their professional research are disposed to conform to Popper's norms.

Thus, in spite of Popper's professed antinaturalism, his methodological rules do have a descriptive or naturalistic facet in the sense that *the justificatory reasons offered for them include empirical claims.* Indeed, Popper's methodology and similar ones have repeatedly been challenged on the empirical ground that, in important contexts, scientists have not conformed to Popper's canons. I will not enter into those criticisms; my concern here is simply to call attention to the *naturalistic facet of Popper's methodology.*

However, that methodology also has a normative or valuational facet. Popper's choice of examples of scientific theorizing is prompted by his view that properly scientific theories *should* be marked by precise testability, large content, and similar features. The norms invoked here are not moral, or ethical, or prudential ones; they might rather be called *epistemological norms or values,* which, in the search for improved knowledge, assign high importance to susceptibility to severe tests and possible falsification.

Both empirical information and epistemological valuation, then, are required for critically appraising or supporting a methodology of the kind intended by Popper. If his principles were presented as arbitrary conventions chosen as the rules of a "game of science" designed by Popper, they would have no epistemological interest whatsoever.

What lends them such interest is that the rules are meant as a partial and assuredly idealized characterization of a peculiar and highly important kind of human enterprise – namely, scientific research. Thus, Popper's methodology, despite its professed anti-naturalism, possesses both the normative and the descriptive or "naturalistic" facets just indicated.

4. Carnap on the Explication of Methodological Concepts

Analogous considerations are applicable, I think, to Carnap's conception of the cognitive status of the philosophy of science. According to Carnap, the principles of an adequate methodology of empirical science are established by a process he calls *explication* (1950: chapter 1). In this process, such concepts as testing, confirmation, falsification, explanation, etc. are analytically clarified and then given refined and precise re-definitions. The idea subjected to such clarification is referred to as the *explicandum*; the new precise concept by which it is replaced is called the *explicatum*. The process has played an important role in analytic empiricism, where it has often been referred to as *logical analysis* or as *rational reconstruction.* The accounts proposed by analytic empiricists for such notions as verifiability, falsifiability, confirmation, inductive probability, theoretical reduction, and modes of explanation are all aimed at explication; they offer explicit and precise reconstructions of vague but important concepts of epistemology and methodology. Carnap requires adequate explications to be stated in logically *precise* ways, to be as *simple* as possible, and to provide a basis for *fruitful* philosophical theorizing (1950: 7).

This view of methodological principles as the results of explicatory re-definitions of methodological concepts is akin, I think, to Popper's conception of methodological principles as established by convention; the principles of a methodological theory based on Carnapian explication would seem to hold in an analytic, a priori fashion by virtue of the relevant explicatory re-definitions of methodological concepts.

Again, however, this is no more true of Carnap's than of Popper's conception, for Carnapian explications are intended to clarify and sharpen certain concepts – such as testability, rational credibility, and so on – *which are already in use and play an important role in the preanalytic discussion of scientific procedures.* Adequate explications must, therefore, conform to a reasonable extent to the preanalytic, vague use of the *explicandum* terms. Carnap, indeed, establishes this as a further requirement, in addition to the conditions of precise formulation, simplicity, and fruitfulness: "The explicatum must be *similar to the explicandum*" (1950: 7), which roughly means that the explicated precise concept must apply to most of those cases to which the vague, preanalytic concept is clearly meant to apply; and it must not apply to those cases to which the preanalytic concept is generally not applied. Thus, the similarity requirement imposes *empirical constraints* on explication while, at the same time, leaving room for a prescriptive conventional component when choosing some particular explicatory re-definition so as best to comply with the requirements of simplicity, precision of formulation, and fruitfulness.

As an example of an explicatory theory of great interest for the methodology of empirical science, consider Carnap's theory of the logical or inductive probability, $p(H,E)$, of a hypothesis H relative to evidence E. By way of a vague preanalytic characterization, Carnap says that the concept is to represent the degree to which a person is rationally entitled to *believe* the hypothesis H on the basis of E and that it is also to be a fair betting quotient for a bet on H for someone whose entire information concerning H is E (1963a: 967).

Carnap's explication is formulated in the context of an axiomatized theory in which $p(H,E)$ is defined in precise terms as a quantitative, purely logical relation between the hypothesis H and the evidence sentence E.

What is the cognitive status of Carnap's theory of rational credibility? Is it purely a matter of convention, or does it have an empirical facet?

Carnap's argument is, briefly, that the axioms of his precise theory of rational credibility reflect our intuitive beliefs about types of bets it would be rational to engage in and the like. Thus, he says, for example, "the reasons to be given for accepting any axiom of inductive logic. . . . are based upon our intuitive judgments. . . . concerning the inductive rationality of practical decisions (e.g., about bets)." He also claims that those "reasons are a priori [i.e.,] independent both of universal synthetic principles about the world . . . and of specific past experiences" (Carnap 1963a: 978–9).

This reliance on our intuitive judgments about the rationality of decisions is, I think, closely akin to Goodman's view concerning the justification of rules of deductive and inductive reasoning. Goodman holds that particular inferences are justified by their conformity with general rules of inference and that general rules are justified by their conformity with valid particular inferences. "The point is," he says, "that rules and particular inferences alike are justified by being brought into agreement with each other. *A rule is amended if it yields an inference we are unwilling to accept; an inference is rejected if it violates a rule we are unwilling to amend*" (1955: 67).

When Carnap speaks of "our" intuitive judgments concerning rationality and when Goodman refers to particular inferences or to general rules "we" are unwilling to accept or to amend, who are "we"? Surely, those intuitions and that unwillingness are not meant to be idiosyncratic; the idea is not "to everyone his own standards of rationality." Carnap surely assumes that there is a body of widely shared intuitions in the matter and that approximate conformity with them provides a justification for accepting certain rules of inductive reasoning. Indeed, otherwise, there would be no *explicandum*, and the problem of an explicatory theory could not arise.[1]

I think, therefore, that the grounds Carnap offers in support of his explicatory theory of rational inductive belief are not just a priori, as he asserts; for they make descriptive socio-

---

[1] For a provocative critical and constructive discussion of Goodman's view and of difficulties in the idea of a widely shared consensus, compare Stich and Nisbett (1980) and the further considerations offered by Conee and Feldman (1983).

psychological claims about shared intuitive judgments concerning the *explicandum* concept. Just as that concept is vague, so, admittedly, are those supporting claims. They do not specify, for example, exactly whose intuitions support the theoretical explication. But if, for example, some of the basic intuitive judgments adduced by Carnap were deemed counterintuitive by a large proportion of scientists, mathematical statisticians, and decision-theorists, then surely they could not serve to justify Carnap's theory itself. Indeed, in a remark on just this point, Carnap acknowledges that, in contrast to his theory, scientists do not as a rule explicitly assign numerical degrees of credibility to hypotheses; but he adds,

> it seems to me that scientists show, in their behavior, implicit use of these numerical values. For example, . . . [a physicist's] practical decisions with respect to his investment of money and effort in a research project show implicitly certain features . . . of his credibility function. . . . If sufficient data about decisions of this kind made by scientists were known, then it would be possible to determine whether a proposed system of inductive logic is in agreement with these decisions (Carnap 1963b: 990).[2]

This passage clearly comes close to claiming that an explicatory theory of rational credibility as conceived by Carnap should not just prescribe norms for rational research procedures but should also have the potential for providing at least an approximate descriptive and explanatory account of some aspects of actual scientific inquiry.

The component of prescription or of epistemological valuation in Carnap's explication of rational credibility becomes manifest in his requirements of precision, simplicity, and fruitfulness – and in certain stipulations he characterizes as general conditions of rationality. Among these is his stipulation that the rational credibility assigned to an hypothesis should be "dependent, not upon irrational factors like wishful or fearful thinking, but only

---

[2] Note also Carnap's interesting remarks (p. 994) about differences and changes in intuitions concerning rational credibility.

on the totality of [the scientist's] observational knowledge at the time" (1963a: 970). This is just a special consequence of the more general epistemic ideal that credibility should be a *purely logical function* of hypothesis and evidence and that psychological factors, therefore, should not enter into it.

Thus, Carnapian explication and the methodological principles it yields have both a prescriptive and a descriptive facet. In this respect, the explication of rational credibility by Carnap's precise theory of inductive probability is quite analogous to the explication of the concept of a correctly formed English sentence by a precise theory of English grammar. Such an explication must surely take account of the linguistic intuitions and dispositions of native speakers; but descriptive faithfulness has to be constrained by and adjusted to the systematic, epistemological value of constructing a grammar which is reasonably precise, simple, and general. Consider, for example, the idea that there must be an upper limit to the length of any correctly formed English sentence. Such an idea has an obvious intuitive appeal and descriptive correctness, but it is logically incompatible with the idea that the conjunction of any two correctly formed English sentences is again a correctly formed English sentence. The latter principle is strongly favored by considerations of theoretical comprehensiveness and simplicity, and these, in addition to the facts of linguistic behavior, have to be accommodated by a satisfactory theory of grammar. Thus – and here lies the analogy to the principles of the methodology of science – the rules of a theory of English grammar are neither purely descriptive of the linguistic behavior of native speakers nor are they simply rules, adopted by convention, for a game we decide to call "speaking English." Theories of methodology, like theories of grammar, have a descriptive as well as a prescriptive facet.

In the preceding discussion, I have been concerned only to show that even methodological theories presented as nonempirical or antinaturalistic, such as Popper's and Carnap's, do have a descriptive, naturalistic component; it was not my purpose to examine the substantive merits or shortcomings of those theories.

I do want to mention, however, that Carnap's explication of

rational credibility as a logical function solely of the hypothesis in question and the available evidence is applicable only to quite limited kinds of hypothesis-testing; it does not do justice to the complex considerations that enter into the critical appraisal of a scientific hypothesis when a background of other hypotheses or theoretical principles is available. The reason lies, briefly, in the Duhem-Quine argument, which has led to a holistic conception of scientific method and knowledge. The point of relevance to Carnap's approach is that experimentally testable statements cannot be derived from the hypothesis under test alone but only from the hypothesis taken in conjunction with an extensive system of other assumptions. Broadly speaking, the relevance of a given body of evidence for the credibility of a hypothesis depends on the entire theoretical system accepted by science at the time.

If experimental findings then conflict with theoretical predictions, the hypothesis may be assigned low credibility and be rejected; but not infrequently, the conflict is resolved in science by making adjustments elsewhere in the total system of accepted hypotheses. What we have so far discussed as the rational appraisal of an individual hypothesis thus comes to be seen as something involving a critical comparison of alternative theoretical systems that may serve to adjust scientific claims and available evidence to each other.

5. *Pragmatist Approaches to Methodology*

This more general problem of how, in the context of scientific inquiry, competing theories are critically compared and of how choices between them are made is not addressed in Carnap's theory of inductive probability. It is, however, one of the principal topics of recent pragmatically oriented methodological studies, among which Kuhn's theory of scientific revolutions has been particularly influential.

Let us, then, look at the historic-sociological perspective on methodological issues and especially at Kuhn's characterization of theory choice in science.

Here the emphasis on the descriptive facet of methodological claims is dominant. An adequate methodological theory must be informed, on this view, by a close study of the history, sociology, and psychology of actual scientific research behavior. A proper descriptive and explanatory account of this kind will then also, it is claimed, illuminate the concept of methodological rationality in scientific inquiry.

Thus, Kuhn argues that, in the pursuit of knowledge, "scientific behavior, taken as a whole, is the best example we have of rationality." Hence,

> if history or any other empirical discipline leads us to believe that the development of science depends essentially on behavior that we have previously thought to be irrational, then we should conclude not that science is irrational, but that our notion of rationality needs adjustment here and there. (Kuhn 1971: 144)

In a similar vein, Kuhn says elsewhere, "If I have a [descriptive] theory of how and why science works, it must necessarily have implications for the way in which scientists should behave if their enterprise is to flourish" (Kuhn 1970b: 237). Hence, in response to Feyerabend's question of whether Kuhn's methodological principles are to be read as descriptions or prescriptions, Kuhn answers, "They should be read in both ways at once" (Kuhn 1970b: 237).

However, the assignment of a prescriptive reading to a descriptive account of scientific research is not quite so straightforward. It presupposes epistemological valuation no less than does the construction of a methodological theory by analytic explication. There are at least two reasons for this.

First, Kuhn's basic assumption that science is the best example we have of rationality expresses a broad epistemological valuation, a judgment as to what *matters* for the rational pursuit of knowledge. This value judgment, which places scientific modes of research highest on the rationality scale, is posited and not further argued. Kuhn seems to suggest just that by remarking that he takes that judgment "not as a matter of fact, but of principle" (1971: 144).

A second way valuation enters Kuhn's argument is this: The behavior of scientists engaged in their professional work admits of many different descriptions. Some of these might mention, for example, the widespread intensive competition among specialists working in the same problem area and the resulting tendency to conceal from each other their guiding ideas, their methods of approach, and their unpublished results. A closely related feature of actual scientific research behavior is the recently much discussed practice of "doctoring" the evidence for scientific claims, or even faking or freely inventing experiments and experimental data in support of such claims. A descriptive study of scientific research would quite properly note such practices, but we would not want to include them in a normative methodology intended to provide standards of rationality in the pursuit of knowledge. Kuhn acknowledges this point implicitly in a remark to the effect that our notion of rationality depends on what we take to be the "essential" aspects of scientific behavior (1971: 144). The term *essential* here surely refers to antecedently assumed – perhaps intuitively espoused – epistemological standards or values.

FRAUD

Thus, the justificatory grounds for methodological theories both of the pragmatist and of the analytic-explicatory varieties have a descriptive or naturalistic component and a component reflecting epistemological prescription or valuation. Neither of the two construals of the methodology of science is purely descriptive, and neither is purely prescriptive and normative.

6. The Character of the Methodological Norms Yielded by Kuhn's Pragmatist Approach

Despite these basic affinities, the character of the prescriptive methodology attainable by Kuhn's pragmatist approach differs significantly from that of the methodological principles proposed by Carnap or Popper.

Analytic explicators aim at formulating explicit and precise and intersubjectively applicable standards for such contexts as

213

the critical appraisal of scientific hypotheses or theories. Carnap bases his standards of appraisal on his rigorously formalized theory of rational credibility; Popper formulates rules for the game of science that are expressed in terms of precise concepts of falsifiability, corroboration, and the like.

Kuhn's pragmatist account, on the other hand, does not envisage – let alone set forth – any precise standards of theory choice. This is readily seen from Kuhn's characterization of the ways scientists appraise competing theories and eventually choose between them. Kuhn's ideas on the subject are well known, and I will mention only a few points that have an immediate bearing on the character of the methodological prescriptions that might be gleaned from his descriptive account.

Kuhn argues that the choice between competing theories is a matter left to the specialists in the field, and he emphasizes that their ways of appraising the merits of such theories are strongly influenced by certain preferences or values which they have in common and which have been shaped in the course of their technical training and professional experiences. In particular, scientists agree widely in giving preference to theories possessing certain characteristics which I will call *desiderata* for short; in the methodological literature, also before Kuhn, such characteristics have often been noted and referred to as *"marks of a good hypothesis."* Among those mentioned by Kuhn are the following: A theory should yield *precise*, preferably quantitative, *predictions*; it should be *accurate* (i.e., consequences derivable from it should be in good agreement with the results of experimental tests); it should be *logically consistent* both internally and with currently accepted theories in neighboring fields; it should *predict phenomena that are novel* in the sense of not having been known or taken into account when the theory was formulated; it should be *simple*; and it should be *fruitful* (1977: 321–2; 1970a: 205–6; 1970b: 245–6).

Kuhn argues, like other thinkers before him, that while scientists are in general agreement about the importance of these characteristics, the *desiderata* cannot be expressed in the form of precise, explicit, unambiguous criteria for the critical comparison of theories for two reasons. First, each individual *desideratum* is

214

too vague to permit a precise explication; the ideal of simplicity is a well-known example. Second, even if precise and generally acceptable criteria could be formulated for comparing competing theories in regard to each of the *desiderata*, there would remain the possibility that one of two competing theories might be superior to the other in regard to some of the *desiderata* but inferior in regard to others. For an overall decision on preferability, a further rule would then be needed which would, so to speak, assign different weights to the different *desiderata*. Again, Kuhn argues plausibly that it is not possible to formulate a general and precise rule of that kind which does reasonable justice to theory choice as actually practiced in science.[3]

Indeed, Kuhn holds that two scientists fully committed to the same list of *desiderata* may nevertheless reach different conclusions in a particular case of theory choice, "without thereby in the least jeopardizing their adherence to the canons that make science scientific" (1977: 324).[4]

Thus, Kuhn's account of scientific theorizing provides no precise explication of those canons; indeed, Kuhn's use of the word "canon" in this context may be slightly misleading. In view of the general considerations just mentioned, I think that Kuhn's position on this issue is correct. A comprehensive, formally precise, and substantively adequate rational reconstruction of standards of scientific theory choice is not to be expected.

Carnap himself acknowledged the difficulty. In regard to his own explicatory theory of rational credibility, he noted, "Inductive logic alone . . . cannot determine the best hypothesis on a given evidence, if the best hypothesis means that which good scientists would prefer. This preference is determined by factors of many different kinds, among them logical, methodological, and purely subjective factors" (1950: 221). He adds, "[H]owever, the task of inductive logic is not to represent all the factors, but only

---

[3] Some similar arguments had been presented earlier by Kries (1886: 26, 28ff) and by Nagel (1939: 68–71); Carnap appraises these arguments in *Logical Foundations of Probability* (1950: 219–33).

[4] The entire essay offers a very suggestive presentation of the ideas referred to here. For other passages concerning these issues, see Kuhn (1970b: 241, 245–6, 261–2; 1970a: 199–200).

the logical ones; the . . . nonlogical factors lie outside its scope"
(1950: 219).

Abstracting from certain aspects of actual scientific procedure
makes it possible, he notes, to construct a precise and powerful
explicatory theory; but he cautions against

> the ever-present temptation to overschematize and
> oversimplify . . . ; the result may be a theory which is
> wonderful to look at in its exactness . . . and yet woefully
> inadequate for the tasks of application for which it is intended.
> (This is a warning directed at the author of this book by his
> critical superego.) (Carnap 1950: 218)

Kuhn and like-minded thinkers of a pragmatist bent agree
with Carnap's superego. Carnap achieved formal power and
elegance at the price of excluding from his theory various
methodological considerations which play an important role in
scientific theorizing but which do not lend themselves to precise
formalization.

However, to acknowledge the importance of certain factors not
amenable to precise explication is not to advocate methodological anarchy in science; for, first of all, there are indeed various
important standards that do admit of precise explication and
are certainly acknowledged by the scientific community. These
include the requirements of consistent and valid logical and
mathematical reasoning when examining the implications of
scientific theories. Similarly, quite precise methodological rules
can be formulated for scientific procedures of limited scope, such
as those used in the context of measurement or the testing of
statistical hypotheses.

Even the important considerations implicit in the *desiderata*,
though not precisely statable, are, in many cases, applicable
with considerable intersubjective agreement. This is so, one may
speculate, because the *desiderata* reflect a profound and widely
shared human concern whose satisfaction is the principal goal of
scientific research – namely, the formation of a general account
of the world which is as accurate, comprehensive, systematic,
and simple as possible and which affords us both understanding
and foresight.

All these features of the intended account of the world are reflected in the *desiderata*. That the desire for such a world view is widely shared and that we can agree, at least roughly, in our appraisal of competing systems are themselves empirical features of the world which might be scientifically explored, perhaps along evolutionary lines. At any rate, in its absence, there would be no science as we know it, and we would not be here to wonder about its methods.

II. EVIDENCE AND TRUTH IN SCIENCE

1. *Introduction*

The ultimate goal of scientific inquiry is often said to be the attainment of truth, the establishment of true theories about the world.

To what extent does this conception accord with the ways in which hypotheses and theories are tested and eventually accepted or rejected in the context of scientific research? I propose to consider this question in light of the characterizations of scientific procedure that have been developed in the tradition of logical or analytic empiricism and, more recently, within a broader pragmatist and naturalistic perspective.

2. *Induction and*
Verification

The basic idea of empiricism is the view that all our knowledge about the world – in particular, scientific knowledge – ultimately rests on or is supported by the data of our immediate experience. This conception is the root of the venerable problem of induction, which may usefully be divided into two components. The first of these is the problem of characterizing, perhaps by precise logical criteria, the relation of support between experiential evidence and empirical claims; the second component is the question of what bearing such evidential support has upon the truth of scientific hypotheses and theories.

As David Hume has shown, individual observational findings,

217

however numerous, cannot in general support an empirical statement in the sense of proving it true or verifying it with deductive certainty. The finding that all events of kind A that have so
far been observed were accompanied by an event of kind B does
not permit a deductive inference to the prediction that the next
occurrence of A will be associated with an occurrence of B, and
it permits even less the deduction of a generalization to the effect
that every occurrence of kind A, whenever and wherever it may
take place, is accompanied by an occurrence of kind B.

In this respect, the nonempirical science of mathematics seems
to be much better off. The theorems of a mathematical theory can
be derived from that theory's axioms with deductive certainty.
But this certainty is bought at a high price. For the content of
the conclusion of a deductive inference never reaches beyond
the content of the premises. Even the most complex theorem of
Euclidean geometry, for example, merely repeats part of what is
already said in the axioms. However surprising the theorems
may be from a *psychological* point of view, *logically* speaking they
contain no new information compared with the axioms. This is
precisely the reason for their certainty relative to the axioms.

Evidence sentences describing observational findings cannot
confer such deductive certainty upon scientific claims. For those
claims go far beyond what has been ascertained by past observations; they assert universal laws and imply statements about
occurrences, both past and future, that have not been observed
and are not included in the available evidence. Insistence on
"proof" by evidence would, therefore, defeat this basic objective
of scientific inquiry.

3. Falsification, Confirmation

In view of these difficulties, Popper developed his influential
falsificationist doctrine that observational findings cannot serve
at all to lend inductive support to scientific statements but only
to test them critically and possibly to *falsify* or *refute* them *deductively*. Thus, the universal statement that the specific volume of
every liquid decreases with decreasing temperature could be

refuted by observational findings according to which the specific volume of experimentally examined samples of water increased when they were cooled from 4°C to 0°C. In Popper's view, the evolution of scientific knowledge appears as a sequence of hypothetical conjectures and experimental refutations.

Falsificationism faces difficulties, however. Hypotheses of universal conditional form are indeed falsifiable by negative instances. But, for purely logical reasons, hypotheses of existential or mixed quantificational form do not admit of such falsification. There is a further difficulty: The falsifiability of a hypothesis presupposes that the occurrence of certain observable events is deducible from the hypothesis *alone*. But as Pierre Duhem showed at the beginning of this century, this presupposition is practically never fulfilled. The deduction of observational findings from a hypothesis usually requires many additional premisses, among them a system of assumptions about the design and mode of functioning of the experimental apparatus. Thus, nonoccurrence of the deduced observational results shows at best that at least one of the premisses presupposed in the deduction is false but not that this must be the hypothesis under test.

The idea suggests itself that although, in general, observational findings can neither strictly *verify* nor strictly *falsify* a hypothesis, they might *confirm* or *support* a hypothesis *more or less strongly*, where the degree of support might be taken as a measure of the *probability* or *rational credibility* conferred upon the hypothesis by the observational data. A view like this was suggestively propounded by John Maynard Keynes in his work on probability. Keynes characterized the relevant notion of probability through a formal system of axioms and investigated its presuppositions and implications for various applications. He did not, however, establish rules by means of which the probability of a hypothesis relative to any given observation reports might be calculated.

Around 1950, Carnap formulated his formalized theory of inductive probability. Its axioms are much simpler than those of Keynes) and, in addition, it provides explicit definitions which assign a definite numerical probability to any given hypothesis relative to any given observational evidence. But Carnap's theory requires that the hypotheses and observational data in question be expressed in a formalized language of a logically simple structure – a presupposition not fulfilled by advanced scientific theories.

4. Pragmatist Perspectives on the Problem of Induction

As considered so far, induction is a procedure for the critical evaluation of hypotheses and theories with regard to their support by available empirical evidence. But this view would be much too narrow even if it should be possible to develop a general theory of degrees of empirical support – for example, in Carnap's sense – which would reach far beyond its present area of application. For in the critical scientific appraisal of a hypothesis or theory, it is not only its empirical support which is taken into consideration but also a whole series of additional factors, which I shall briefly call *desiderata* because they count as desired characteristics of hypotheses and theories. Such additional criteria of appraisal have frequently been mentioned in the methodological literature; they play an especially important role in Kuhn's theory of scientific revolutions and in other pragmatist perspectives on scientific inquiry.

One of these *desiderata* calls for an optimally precise, logically consistent formulation of a theory, one permitting the derivation of empirically testable consequences; in addition, it calls for good agreement of these consequences with observational or experimental findings. I shall call this requirement *the basic empiricist desideratum*. Another desirable property of a theory is wide *scope* in the sense that the theory covers phenomena of many different kinds and predicts novel events. *Compatibility* of a theory with *theories presently accepted in neighboring disciplines* counts as important, and a further *desideratum* often mentioned is *simplicity*.

Logical – or analytic – empiricism would see the task of

The Status and Rationale of Scientific Methodology

methodology as formulating an *explication* or *rational reconstruction* which characterizes the various *desiderata* and their roles in the critical evaluation of theories by means of objectively applicable concepts and rules. One might even envisage the ideal of expressing the extent to which a theory fulfills the various *desiderata* by means of a combined quantitative "theoretical utility value" of the theory, or one might think of establishing objective criteria for a nonquantitative comparison of theories with regard to the fulfillment of the *totality* of the *desiderata*.

But such projects are utopian. With few exceptions, the *desiderata* are too vague to permit a precise explication of theory choice which provides, on the one hand, sharp decisions between competing theories and which does sufficient justice, on the other hand, to the character of scientific research. The requirement of logical consistency can indeed be precisely expressed, and, to a limited extent, so can the basic empiricist *desideratum* of the derivability of empirically testable consequences which accord with experimental findings. But nothing similar can be expected for the notion of the scope of theories, for the degree of novelty of their predictions, and for their compatibility with theories presently accepted in neighboring fields, not to mention their simplicity – a *desideratum* which comprises diverse ideas, such as the ease with which the theory's mathematical apparatus can be handled and the perspicuity and elegance of the basic theoretical concepts and assumptions.

Undeniably, such *desiderata* exert a far-reaching influence on the appraisal and adoption of scientific hypotheses and theories, but they cannot, in their totality, be expressed by precise rules of inductive appraisal. This does not mean, however, that choice of scientific theories is a subjective, perhaps altogether arbitrary, procedure. In this context, Kuhn's ideas seem to me helpful and enlightening. Kuhn emphasizes that choosing between competing theories is a task left to the specialists in the field and that these specialists share certain standards – the *desiderata* – which they acquire and refine in the course of their training and professional experience, without learning or obeying an explicitly formulated system of rules. What they acquire in the course of their professional training are decisional dispositions essential for their professional judgments but which cannot be character- ized as dispositions to obey certain explicit criteria or rules.

These considerations illustrate the recent turn from a purely logico-analytic, antinaturalistic characterization of scientific method to a more comprehensive conception which includes naturalistic-pragmatist – particularly socio-psychological consid- erations – without thereby making theory choice a matter of mass psychology.

5. Desiderata and *Truth*

The preceding considerations raise a question about the view that it is the ultimate goal of science to ascertain the truth about the world. Is the scientific way of appraising and choosing theo- ries indeed conducive to the attainment of truths? Is it a rational means in the pursuit of that end? What systematic relevance does conformity with the *desiderata* possess for the question of truth?

The *desideratum* of logical consistency clearly is a necessary – though by no means sufficient – condition for the truth of a theory.

The basic empiricist *desideratum* of good agreement between experimentally testable implications of a theory and reports on experimental results might similarly seem to be a necessary con- dition for the truth of the theory. But this idea presupposes that scientific observation reports are *true* descriptions of the observed occurrences, a presupposition that is untenable. Any report on observational findings, instrument readings, or other experimental results is subject to the risk of errors of observation and other disturbing factors. To compare theoretically derived predictions with evidence reports on observational or experi- mental findings is not to compare them with indubitably estab- lished truths.

Moreover, even in the strictest disciplines, reports about empirical findings are sometimes rejected or disregarded because they conflict with currently accepted theories that have so far been very successful. A comprehensive theory of that kind is not

abandoned because it conflicts with a few recalcitrant observation reports – especially when no better theory is in sight. This
methodological peculiarity is clearly irreconcilable with the idea
that all reports on observational and experimental findings are
true, and it questions the view that the modes of scientific
theory choice have a systematic bearing on the attainment of
truth.

As for the *desideratum* of large scope, it surely is no indication
of truth. In particular, when one theory has a wider scope than
another in the strict sense that it entails the latter without being
entailed by it, then the stronger, more far-reaching theory has a
smaller probability of being true than does the weaker, less far-
reaching one.

What about the *desideratum* of simplicity? If the simplicity of
a theory is taken to lie in the ease with which its mathematical
apparatus can be handled, then the desideratum clearly has no
systematic relevance for the question of truth. If the *desideratum*
is meant to refer to the conceptual economy and perspicuity of
the basic theoretical ideas, it may reflect the metaphysical idea
that the world itself has a simple fundamental structure and that,
therefore, the simpler theory will *ceteris paribus* come closer to
the truth. In a similar view, the theoretical physicist Paul Dirac
expressed the conviction that the true laws of nature are mathematically beautiful because God made the universe and gave it
a beautiful structure. Indeed, he suggested by way of heuristic
advice that theoretical physicists should search for theories that
are mathematically beautiful. Such general conceptions have no
doubt inspired many a great scientist; but they establish no objective link between simplicity (or beauty) and truth.

It might seem that the issue of the relevance of the *desiderata*
for the truths of scientific claims could be entirely avoided by a
structuralist ("nonstatement") construal of scientific theories as
propounded by P. Suppes, J. Sneed, W. Stegmuller, and others.
Roughly speaking, that construal presents theories not as classes
of statements but as deductively organized classes of sentential
functions, which are neither true nor false. The empirical role of
such systems is indicated by ascribing empirical models to them.
For example, the solar system might be claimed to be a model of

C 13

some structuralist formulation of Newton's theory of motion and gravitation. But the assertion that a specified physical system constitutes a model for a given structuralistically formulated theory is an empirical assertion for which the question of truth arises. Thus, the structuralist mode of characterizing scientific theories, while illuminating in many respects, does not offer a way of entirely avoiding the question of truth in the content of scientific theorizing; indeed, the structuralist talk of a "nonstatement view" of scientific theories is not meant to have that implication.

Skeptical considerations analogous to those outlined before can be adduced to show for other *desiderata* that they have no logical bearing on the truth of the theories appraised and accepted in conformity with them; adherence to them does not mark scientific inquiry as a rational means for the pursuit of truth.

Instead, scientific research might be said to aim at attaining theories that are *epistemically optimal* in a sense roughly indicated, though by no means precisely explicated, by the *desiderata* (i.e., theories which, at the current stage of inquiry, are as accurate, precise, and comprehensive as possible, which provide explanations and predictions of experimental findings, and so on). Relative to this goal, it is obviously – indeed, trivially – rational to choose and to change scientific hypotheses and theories in accordance with the *desiderata*.

Should we then, perhaps, entirely abandon the classical concept of truth referred to so far and replace it with a radically pragmatist construal of truth as epistemic optimality? After all, is not critical appraisal in the light of the *desiderata* the method par excellence by which science adjudges the merits of empirical claims?

In the traditions both of logical empiricism and of pragmatism, it has indeed sometimes been argued that all we can ever ascertain and all that matters for human purposes about scientific theories are whether they agree with the available evidence and offer effective means of organizing our experiences and predicting new ones. Hence, this view holds that the classical conception of truth as agreement with what is the case should be replaced by a construal of truth as, broadly speaking, scientific acceptability or, as we might put it, epistemic optimality. This switch would salvage the characterization of science as a search for truth, but at the price of an ad hoc re-definition that confounds the important but quite different roles played by the classical semantical concept of truth and the epistemological concept of acceptability or epistemic optimality.

The classical concept of truth involves no reference at all to evidence or *desiderata*. A sentence in a given language is either true or false, independent of whether it is ever tested. Its truth-value does not change with changes in the available evidence. Epistemic optimality, on the other hand, is a relational concept: It can be attributed to a theory only by comparison with specified competing alternatives and only relative to a given "knowledge situation," including the available evidence, currently accepted background assumptions, and other information pertinent to the satisfaction of the *desiderata*. Truth satisfies the principle of excluded middle; epistemic optimality does not. It is not the case that a theory is epistemically optimal (in a given knowledge situation) if and only if its negation is not.

The classical concept of truth has this basic characteristic: A descriptive sentence S is equivalent to the sentence "S is true"; to say that S is true is tantamount to asserting S. Hence, if in a given knowledge situation, a theory T_1 is epistemically superior to a theory T_2, then, in that knowledge situation, the claim "T_1 is true" is epistemically superior to the claim "T_2 is true." But since epistemic superiority has no logical link to truth, this does not mean that T_1 "comes closer to the truth" than T_2. Particularly, truth cannot be represented as a limiting case or ideal case of epistemic superiority.

6. *The Status of Evidence*

After this sketchy survey of the role of *desiderata* and their bearing on truth, let us briefly return to what I called the basic empiricist *desideratum*. This *desideratum* reflects the conception that scientific claims, to be acceptable, must be in accord with experiential evidence. And surely, this requirement counts as essential in scientific inquiry. The proponents of even the most speculative theories about the origins and evolution of the universe, for example, agree and indeed insist that their ideas must eventually be amenable to empirical test, that they must lead to predictions that can be checked by comparison with observational evidence. As noted earlier, however, agreement with observational reports does not guarantee the truth of the predictions in question since test reports can be distorted by various factors, ranging from physiological and psychological conditions to negligence and outright fraud. Even if the testing is not done by human observers but, say, by instruments producing "observation reports" in the form of automatic printouts, the resulting evidence sentences always remain open to reconsideration and revision. The acceptability of evidence sentences produced by human or instrumental observers is itself an empirical matter, one open to scientific investigation in light of available knowledge about the structure and mode of functioning of the observational system in question. No observation report is acceptable with finality; any accepted evidence is, in principle, defeasible.

In this considerably relaxed empiricist perspective, does it make sense at all to apply the concept of truth to scientific theories? If truth is understood in the semantical sense, the answer is in the affirmative, for to assert that a theory is true is tantamount to asserting the theory itself.

What has to be rejected is, rather, the idea that scientific inquiry informed by experiential evidence might give us definitive *knowledge* of the truth of at least some empirical beliefs. In accordance with a holistically and pragmatically liberalized empiricism, the bearing of given evidence on scientific claims and thus on their truth is subject to appraisal and remains open to reappraisal in light of pertinent theories accepted as epistemically optimal at the time.

REFERENCES

Carnap, Rudolf
1950: *Logical Foundation of Probability* (Chicago: University of Chicago Press).

1963a: "My Basic Conceptions of Probability and Induction," in Paul A. Schlipp, ed., *The Philosophy of Rudolf Carnap* (La Salle, IL: Open Court), pp. 966–79.

1963b: "Ernest Nagel on Induction," in Paul A. Schlipp, ed., *The Philosophy of Rudolf Carnap* (La Salle, IL: Open Court), pp. 989–95.

Conee, Earl, and Richard Feldman

1983: "Stich and Nisbett on Justifying Inference Rules," *Philosophy of Science* 50, 326–31.

Goodman, N.

1955: *Fact, Fiction, and Forecast* (Cambridge, MA: Harvard University Press).

Hempel, Carl G.

1943: "A Purely Syntactical Definition of Confirmation," *Journal of Symbolic Logic* 8, 122–43.

1945: "Studies in the Logic of Confirmation," *Mind* 54, 1–26, 97–121.

Kries, J. von

1886: *Die Prinzipien der Wahrscheinlichkeitsrechnung* (Freiburg: J. C. Mohr; Türbingen: J. C. Mohr, 1927).

Kuhn, Thomas S.

1970a: *The Structure of Scientific Revolutions* (Chicago: University of Chicago Press).

1970b: "Reflections on My Critics," in I. Lakatos and A. Musgrave, eds., *Criticism and the Growth of Knowledge* (New York: Cambridge University Press).

1971: "Notes on Lakatos," in Roger C. Buck and Robert S. Cohen, eds., *PSA 1970, Proceedings of the 1970 Biennial Meeting, Philosophy of Science Association* (Hingham, MA: D. Reidel Publishing Company), pp. 137–46.

1977: "Objectivity, Value Judgement, and Theory Choice," in T. S. Kuhn, *The Essential Tension* (Chicago: University of Chicago Press), 320–39.

Nagel, Ernest

1939: *Principles of the Theory of Probability* (Chicago: University of Chicago Press).

Popper, Karl R.

1959: *The Logic of Scientific Discovery* (London: Hutchinson).

1962: *Conjectures and Refutations* (New York/London: Basic Books).

Methodology

1979: *Objective Knowledge* (Oxford: Clarendon Press).
Schlipp, Paul A., ed.
1963: *The Philosophy of Rudolf Carnap* (La Salle, IL: Open Court).
Stich, Stephen P., and Richard E. Nisbett
1980: "Justification and the Psychology of Human Reasoning,"
Philosophy of Science 47, 188–202.

Chapter 12

Provisoes

A Problem Concerning the Inferential Function
of Scientific Theories*[1]

1. INTRODUCTION

The principal goal and the proudest achievement of scientific
inquiry is the construction of comprehensive theories which give
us an understanding of large classes of empirical phenomena and
enable us to predict, to retrodict, and to explain them.

These various functions of theories are usually regarded as
having the character of inferences which lead, by way of theoretical principles, from sentences expressing initial and boundary conditions to statements describing the occurrences to be
predicted, retrodicted, or explained.

In this paper, I propose to examine a basic difficulty which
faces this inferential construal of scientific theorizing and which
has implications for some central issues in the philosophy of

* From Carl G. Hempel, "Provisoes: A Problem Concerning the Inferential Function of Scientific Theories," *Enkenntnis* 28 (1988), 147–64. Reprinted with the
kind permission of Kluwer Academic Publishers.
[1] This article has grown out of a paper read in November 1980 at a workshop
held under the auspices of the Center for Philosophy of Science at the University of Pittsburgh. The present, much revised version was written for inclusion in a volume, to be published by the University of California Press, which
is to contain the proceedings of that workshop.

Pending the completion of that project, which has been considerably
delayed, the article appears here with the consent of the editors of the proceedings.

This paper is based upon work supported by National Science Foundation
Grant No. SES 80–25399.

science. I will first present the problem by reference to a purely deductivist conception of theoretical reasoning and will then broaden its scope.

2. THE STANDARD DEDUCTIVIST MODEL

The best-known precise elaboration of a deductivist conception is provided by the so-called standard empiricist construal of theories and their application. It views a theory T as characterizable by an ordered pair consisting of a set C containing the basic principles of the theory and a set I of interpretative statements:

$$(1) \qquad T = \langle C, I \rangle$$

The sentences, or formulas, of C serve to characterize the specific entities and processes posited by the theory (e.g., elementary particles and their interactions) and to state the basic laws to which they are assumed to conform. These sentences are formulated with the help of a theoretical vocabulary, V_C, whose terms refer to the kinds and characteristics of the theoretical entities and processes in question.

The sentences of the interpretative set I serve to link the theoretical scenario represented by C to the empirical phenomena to which the theory is to be applied. These phenomena are taken to be formulated in a vocabulary V_A which is antecedently understood (i.e., which is available and understood independently of the theory). Thus, the sentences of I are said to provide partial interpretations, though not necessarily full definitions, of the theoretical terms in V_C by means of the antecedently understood terms of V_A. So-called operational definitions and reduction sentences in Carnap's sense may be viewed as special kinds of interpretative sentences.

By way of a simple example, assume that T is an elementary theory of magnetism whose theoretical vocabulary V_C contains such terms as "magnet," "north pole," and "south pole," and whose theoretical principles include the laws of magnetic attraction and repulsion and the law that the parts of a magnet are magnets again, while the class I includes some operational criteria for the terms of V_C.

Consider now the following application of the theory. From the sentence "b is a metal bar to which iron filings are clinging" (S_A^1), by means of a suitable operational criterion contained in the set I, infer "b is a magnet" (S_C^1); then, by way of theoretical principles in C, infer "If b is broken into two bars (b_1 and b_2), then both are magnets and their poles will attract or repel each other" (S_C^2). Finally, using further operational criteria from I, derive the sentence "If b is broken into two shorter bars and these are suspended by long thin threads close to each other at the same distance from the ground, they will orient themselves so as to fall into a straight line" (S_A^2). (Note that V_A is here taken to contain not only predicates like "metal bar" but also individual constants such as "b.")

The basic structure thus attributed to a theoretical inference is suggested by the following schema, in which the notation $P \overset{Q}{\rightarrow} R$ is to indicate that R can be inferred from P by using sentences from Q as additional premisses.

(2)

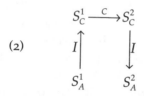

Thus, if the inferential steps in question are indeed all deductive, then the theory provides a deductive inference bridge leading from one V_A-sentence, through the theoretical realm of C, to another V_A-sentence. More precisely: S_A^1 in combination with the theory T deductively implies S_A^2; this, in turn, is tantamount to saying that T deductively implies a corresponding V_A-sentence – namely, the conditional $S_A^1 \supset S_A^2$.

Carnap and other logical empiricists assumed that the vocabulary V_A, which serves to describe the phenomena to be explained by the theory, consists of terms that are "observational" at least in a broad sense (i.e., that they refer to features of the world whose presence or absence can be established by means of more or less direct observation). In recognition of the difficulties that face the notion of observability, I want to avoid any such assumption here. Indeed, I want to provide specifically for cases in which, as often happens, the vocabulary V_A was originally introduced in the context of an earlier theory. All that the standard construal needs to assume is that the phenomena for which the theory is to account are described by means of a vocabulary V_A that is "antecedently available" in the sense that it is well understood and is used with high intersubjective agreement by the scientists in the field. The interpretative sentences in I may then be viewed as interpreting the new terms introduced by the theory (i.e., those in V_C, by means of the antecedently understood terms in V_A).

This deductivist construal[2] faces two basic difficulties. I will call them the problem of theoretical ascent and the problem of provisoes. Let me spell them out in turn.

3. THEORETICAL OR INDUCTIVE ASCENT

The first inferential step in the schematic argument about the bar magnet presupposes that with the help of interpretative sentences belonging to Part I of the theory of magnetism, S_C^1 is deducible from S_A^1. Actually, however, the theory of magnetism surely contains no general principle to the effect that when iron filings cling to a metal bar, then the bar is a magnet. The theory does not preclude the possibility, for example, that the bar is made of lead and is covered with an adhesive to which the filings stick, or that the filings are held in place by a magnet hidden under a wooden board supporting the lead bar. Thus, the theory does not warrant a deductive transition from S_A^1 to S_C^1. It is more plausible to assume that the theory contains an interpretative principle which is the converse of the one just considered – namely, that if a bar is a magnet, then iron filings will cling to it. But even this is not strictly correct, as will be argued shortly.

Hence, the transition from S_A^1 to S_C^1 is not deductive even if the

---

[2] I have limited myself here to a schematic account of those features of the empiricist model which are of relevance of the problems subsequently to be discussed. For fuller expositions and critical discussions – and for references to the extensive literature – see, for example, Carnap (1956, 1966, chaps. 23–6); Feigl (1970); Hempel (1958, 1969, 1970); Putnam (1962); and Suppe (1974), whose comprehensive study includes a large bibliography.

entire theory of magnetism is used as an additional premise. Rather, the transition involves what I will call *inductive of theoretical ascent* (i.e., a transition from a data sentence expressed in V_A to a theoretical hypothesis S_C^1 which, by way of the theory of magnetism, would explain what the data sentences describe).

This illustrates one of the two problems mentioned before that face a strictly deductivist construal of the systematic connections which a theory establishes between V_A-sentences (i.e., between sentences describing empirical phenomena in terms of V_A). This problem has been widely discussed, and various efforts have been made to resolve it by constructing theories of inductive reasoning that would govern such theoretical ascent. I will not consider those efforts here. I will rather turn to the problem of provisoes, which has not, it seems to me, been investigated in the same detail.

4. PROVISOES

Consider the third step in our example – the transition from S_C^2 to S_A^2. Again, the theory of magnetism does not provide interpretative hypotheses which would turn this into a strictly deductive inference. The theory clearly allows for the possibility that two bar magnets, suspended by fine threads close to each other at the same level, will not arrange themselves in a straight line. For example, if a strong magnetic field of suitable direction should be present in addition, then the bars would orient themselves so as to be parallel to each other; similarly, a strong air current would foil the prediction; and so forth.

The theory of magnetism does not guarantee the absence of such disturbing factors. Hence, the inference from S_C^2 to S_A^2 presupposes the additional assumption that the suspended pieces are subject to no disturbing influence or – to put it positively – that their rotational motions are subject only to the magnetic forces they exert upon each other.

Incidentally, the explanatory inference mentioned a moment ago – from S_C^1 to S_A^1 – presupposes an analogous tacit premiss and thus is not deductive.

I will use the term *provisoes* to refer to *assumptions* of the kind just illustrated, *which are essential, but generally unstated, presuppositions of theoretical inferences.*

Provisoes are presupposed also in ostensibly deductive inferences that lead from one V_C-sentence to another. This holds, for example, in the inference from S_C^1 to S_C^2 in the case of the magnet, for if the breaking of the magnet takes place at a high temperature, the pieces may become demagnetized.

Or consider the application of the Newtonian theory of gravitation and of motion to a system of physical bodies like our solar system. In predicting, from a specification of the state of the system at a time t_0, subsequent changes of state, one finds that the basic idea is that the force acting on any one of the bodies is the vector sum of the gravitational forces exerted on it by the other bodies in accordance with the law of gravitation. That force then determines, via the second law of motion ($f = ma$), the resulting change of velocity and of position for the given body. But the quantity f in the second law is understood to be the *total* force acting on the given body; and the envisaged application of the theory, therefore, presupposes a proviso to the effect that the constituent bodies of the system are subject to no forces other than their mutual gravitational attraction. This proviso precludes not only gravitational forces that might be exerted by bodies outside the system but also any electric, magnetic, frictional, or other forces to which the bodies in the system might be subject.

The absence of such forces is not, of course, vouchsafed by the principles of Newton's theory, and it is for this reason that the proviso is needed.

5. ESCAPE BY INTERPRETATIONS OF PROBABILISTIC FORM?

The foregoing considerations show in particular that when a theory contains interpretative sentences in the form of explicit definitions or of Carnapian reduction chains based on the antecedent vocabulary, the applicability of these sentences is usually subject to the fulfillment of provisoes; they cannot be regarded as unequivocal complete or partial criteria of applicability for theoretical expressions.

This thought might suggest a construal of the interpretative sentences of a theory as expressing only probabilistic rather than strictly general connections between theoretically described states or events and certain associated manifestations, or indicator phenomena, described in antecedently available terms. Such a construal might seem to come closer to scientific usage and at the same time to obviate the need for provisoes; for with probabilistic interpretation sentences, a theory would establish at best probabilistic connections between V_A-sentences. And what would otherwise appear as occasional violations of provisoes would be automatically anticipated by the merely probabilistic character of the theoretical inferences.

Interpretative sentences of probabilistic form have in fact been envisaged by several writers. Carnap did so already in his 1956 essay "The Methodological Character of Theoretical Concepts," which is, I think, his earliest full presentation of the standard empiricist construal of theories. He argues there that many terms functioning in scientific theories cannot be regarded as linked to antecedent terms ("observational terms") by interpretative sentences ("rules of correspondence") of strictly universal form. For such sentences would specify strictly necessary or sufficient observational conditions of applicability for the theoretical terms, whereas scientists, Carnap argues, will treat such conditions not as strictly binding, but as qualified by an "escape clause" to the effect that the observational criteria hold "unless there are disturbing factors" or "provided the environment is in a normal state."[3] Such escape clauses clearly have the character of provisoes in the sense adumbrated earlier. Carnap views them as probabilistic qualifiers functioning in interpretative sentences as theoretical terms. These sentences would state probabilistic rather than strictly necessary or sufficient connections between theoretical expressions and V_A-sentences. Indeed, while Carnap countenances dispositional terms, those linked to V_A by strict reduction chains, he suggests that the terms characteristic of scientific theories have only probabilistic links to the observational basis.[4]

---

[3] Carnap (1956), p. 69.   [4] Compare Carnap (1956), pp. 49, 72.

But while Carnap thus explicitly eschews a purely deductivist construal of the inferential function of theories, he does not specify the form of the probabilistic interpretative sentences he envisages. Indeed, in response to a proposal by Pap[5] concerning probabilistic reduction sentences, Carnap remarks, "[I]t seems to me that for the time being the problem of the best form for [interpretative sentences] has not yet been sufficiently clarified."[6]

However that may be, a probabilistic construal of provisoes faces the difficulty that scientific theories do not, in general, provide probabilistic laws that would obviate the need for provisoes.

Consider, for example, the interpretative sentences that would be required for the term "magnet." They would have to take this form: "In cases where iron filings stick to a metal bar, the probability of the bar being a magnet is p_1." Or, for inferences in the opposite direction, they would have to take this form: "Given that a metal bar is magnetic, the probability that iron filings will cling to it is p_2." But surely, the theory of magnetism contains no sentences of this kind; it is a matter quite beyond its scope to state how frequently air currents, disturbing further magnetic fields, or other factors will interfere with the effect in question. It seems to me that no scientific theory provides probabilistic interpretative statements of this sort, whose application is not itself subject to provisoes.

The same basic consideration applies also, I think, where no well-developed and sharply formulated theories are available – for example, probabilification cannot avoid the need for provisoes in the application of theoretical sentences linking psychological states or events to their behavioral manifestations.

6. SOME CONSEQUENCES OF THE NEED FOR PROVISOES

The conclusion that a scientific theory even of nonprobabilistic form does not, in general, establish deductive bridges between

---

[5] Pap (1963), section II.   [6] Carnap (1963), p. 950.

V_A-sentences has significant consequences for other issues in the philosophy of science.

I will briefly indicate four of these: (a) the idea of falsifiability, (b) the significance of so-called elimination programs for theoretical terms, (c) the instrumentalist construal of scientific theories, and (d) the idea of "the empirical content" of a theory.

A. Falsifiability

One obvious consequence of the need for provisoes is that normally a theory is not falsifiable by V_A-sentences; otherwise, it would deductively imply the negations of the falsifying V_A-sentences, which again are V_A-sentences.

This consideration differs from the Duhem-Quine argument that individual hypotheses cannot be falsified by experiential findings because the deduction from the hypothesis of falsifying V_A-sentences requires an extensive system of background hypotheses as additional premises, so that typically only a comprehensive set of hypotheses will entail or contradict V_A-sentences. The argument from provisoes leads rather to the stronger conclusion that even a comprehensive system of hypotheses or theoretical principles will not entail any V_A-sentences because the requisite deduction is subject to provisoes.

Note that a proviso as here understood is not a clause that can be attached to a theory as a whole and can vouchsafe its deductive potency by asserting that in all particular situations to which the theory is applied, disturbing factors are absent. Rather, a proviso has to be conceived as a clause which pertains to some particular application of a given theory and which asserts that in the case at hand, no effective factors are present other than those explicitly taken into account.

B. Elimination Programs
for Theoretical Terms

The need for provisoes also has a bearing on the so-called elimination programs for theoretical terms. These programs are of particular significance for philosophical qualms about the use, in scientific theories, of terms that are not explicitly defined by means of an antecedently understood vocabulary. The ingenious and logically impeccable methods designed by Ramsey and by Craig[7] circumvent these qualms by showing that the use of theoretical expressions can always be avoided in the following sense: If a theory T consisting of two sentence classes C and I as characterized earlier does yield deductive connections between certain V_A-sentences, then it is possible to formulate a corresponding theory (class of sentences) T_A such that

    i. T_A is expressed in terms of V_A alone.
    ii. T_A is logically implied by T.
    iii. T_A entails "$S_A^1 \supset S_A^2$" (and in this sense establishes a deductive bridge from S_A^1 to S_A^2) if and only if T entails "$S_A^1 \supset S_A^2$."[8]

If the function of a theory is taken to consist in establishing deductive bridges among V_A-sentences, then the theory T_A, which avoids the use of theoretical terms, might be called functionally equivalent to the theory T. This result might suggest the reassuring conclusion that, first, in principle, the use of theoretical expressions can always be avoided without any change in the "empirical content" of a theory as it is expressed by the class of V_A-sentences deducible from it and that, second, talk in terms of theoretical expressions is just a convenient *façon de parler* about matters that are fully expressible in the antecedently understood vocabulary V_A. Analogously, it may seem that all the problems about theoretical ascent and provisoes simply disappear if T is replaced by its functional equivalent T_A.

This impression is illusory, however. For a theory T_A constructed from T in the manner of Ramsey or of Craig yields deductive connections between V_A-sentences if and only if T yields such connections; and scientific theories do not, in general, satisfy this condition. The need for provisoes precludes the

---

[7] For details, see Ramsey (1931, "Theories"); Carnap (1966, "The Ramsey Sentence," chap. 26); Craig (1956); Putnam (1965); Hempel (1965, pp. 210–17).

[8] The theory T_A obtainable by Ramsey's method is quite different, in other respects, from that generated by Craig's procedure. But the differences are irrelevant to the point here under discussion.

general avoidability of theoretical expressions by those elimina-tion methods.

The verdict does not hold, however, if the provisoes qualify-ing the inferential applications of a theory are themselves expressible in the antecedent vocabulary. For if P_A is such a proviso governing the transition, by means of T, from S_A^1 to S_A^2, then T entails the sentence $(P_A \cdot S_A^1) \supset S_A^2$ and thus establishes a deductive bridge between two V_A-sentences.

But it seems that, in general, the requisite provisoes cannot be expressed in terms of V_A alone. In the case of the theory of mag-netism referred to earlier, the provisoes may assert, for example, the absence of other magnetic fields, or of disturbing forces, etc., and they will then require at least the use of terms from V_C in their formulation.

C. Provisoes and the
Instrumentalist Perspective

The preceding considerations analogously cast some doubt on the instrumentalist conception of theories as purely inferential devices which, from an input in the form of V_A-sentences, gen-erate an output of other V_A-sentences, for the need for provisoes shows that theories do not render this service. In each particular case, the applicability of the theoretical instrument would be subject to the condition that the pertinent provisoes are fulfilled; and the assertion that they are fulfilled could not just be added to the input into the theoretical calculating machine, for that assertion would not generally be expressible in V_A.

Thus, if a theory is to be thought of as a calculating instrument that generates new V_A-sentences from given ones, then it must be conceived as supplemented by an instruction manual speci-fying that the instrument should be used only in cases in which certain provisoes are satisfied. But the formulation of those pro-visoes will make use of V_C and perhaps even of terms not con-tained in V_C. Thus, one has to check whether certain empirical conditions not expressible in V_A are satisfied, and that surely pro-vides a tug away from instrumentalism and in the direction of realism concerning theoretical entities.

*D. Provisoes and "The
Empirical Content" of
a Theory*

Similar questions arise in regard to the notion of the experiential "cash value" or "empirical content" of a theory as represented by the set of all V_A-sentences entailed by the theory.

Note first, and incidentally, that thus construed, the empirical content of a theory is relative to the vocabulary V_A that counts as antecedently available, so that one would properly have to speak, not of "the" empirical content of T, but of the V_A-content of T.

But the point here to be made is rather that, usually, a theory does not entail V_A-sentences, and thus the proposed construal of empirical content misfires.

To be sure, there are some deductive theoretical inferences that presuppose no provisoes – for example, the inference, mediated by the law of gravitation, from a sentence S^1 specifying the masses and the distance of two bodies to a sentence S^2 specifying the gravitational attraction that the bodies exert upon each other.

But the further theoretical inference from S^2 to a sentence S^3 specifying the accelerations the bodies will undergo requires a proviso to the effect that no other forces act upon the bodies. If S^2 and S^3 are represented as theoretical sentences, then we have here an example of the need for provisoes not only in establishing theoretical inference bridges between V_A-sentences and V_C-sentences but also in building such bridges between sentences expressed solely in terms of V_C. We will shortly return to this point.

7. FURTHER THOUGHTS ON THE CHARACTER OF PROVISOES

How might the notion of proviso be further illuminated? To say that provisoes are just *ceteris paribus* clauses is unhelpful, for the idea of a *ceteris paribus* clause is itself vague and elusive. "Other things being equal, such-and-so is the case." What other things, and equal to what? How is the clause to function in theoretical reasoning?

Provisoes might rather be viewed as *assumptions of completeness*. The proviso required for a theoretical inference from one sentence, S^1, to another, S^2, asserts, broadly speaking, that in a given case (e.g., in that of the metal bar considered earlier), no factors other than those specified in S^1 are present which could affect the event described by S^2.

For example, in the application of Newtonian theory to a double star, it is presupposed that the components of the system are subject to no forces other than their mutual gravitational attraction and, hence, that the specification given in S^1 of the initial and boundary conditions which determine that gravitational attraction is a complete or exhaustive specification of all the forces affecting the components of the system.

Such completeness is of a special kind. It differs sharply, for example, from that invoked in the requirement of complete or total evidence. This is an epistemological condition to the effect that in a probabilistic inference concerning, say, a future occurrence, the total body of evidence available at the time must be chosen as the evidential basis for the inference.[9]

A proviso, on the other hand, calls not for epistemic but for ontic completeness: The specifics expressed by S^1 must include not all the information available at the time (information which may well include false items), but rather all the factors present in the given case which, in fact, affect the outcome to be predicted by the theoretical inference. The factors in question might be said to be those which are "nomically relevant" to the outcome (i.e., those on which the outcome depends in virtue of nomic connections).

Consider once again the use of Newtonian theory to deduce, from a specification S^1 of the state of a binary star system at time t_1, a specification S^2 of its state at t_2. Let us suppose, for simplicity, that S^1 and S^2 are couched in the language of the theory; this enables us to leave on one side the problem of the inductive ascent from astronomical observation data to a theoretical re-

---

[9] Compare Carnap (1950), pp. 211–13; 494.

description in terms of the masses, positions, and velocities of the two objects.

The theoretical inference might then be schematized thus:

$$(3) \qquad (P \cdot S^1 \cdot T) \rightarrow S^2$$

where P is a proviso to the effect that apart from the circumstances specified in S^1 the two bodies are, between t_1 and t_2, subject to no influences from within or from outside the system that would affect their motions. The proviso must thus imply the absence, in the case at hand, of electric, magnetic, and frictional forces, of radiation pressure, and of any telekinetic, angelic, or diabolic influences.

One may well wonder whether this proviso can at all be expressed in the language of celestial mechanics – or even in the combined languages of mechanics and other physical theories. At any rate, neither singly nor jointly do those theories assert that forces of the kinds they deal with are the only kinds by which the motion of a physical body can be affected. A scientific theory propounds an account of certain kinds of empirical phenomena, but it does not pronounce on what other kinds there are. The theory of gravitation neither asserts nor denies the existence of nongravitational forces, and it offers no means of characterizing or distinguishing them.

It might seem, therefore, that the formulation of the proviso transcends the conceptual resources of the theory whose deductive applicability it is to secure. That, however, is not the case in the example at hand. For in Newton's second law, $f = ma$, "f" stands for the *total* force impressed on the body; and our proviso can, therefore, be expressed by asserting that the total force acting on each of the two bodies equals the gravitational force exerted upon it by the other body; and the latter force is determined by the law of gravitation.[10]

But the application of the theory to particular cases is clearly subject again to provisoes to the effect that in computing the total force, all relevant influences affecting the bodies concerned have been taken into account.

---

[10] I am indebted to Michael Friedman for having pointed this out to me.

When the application of a theory to empirical subject matter is schematically represented in the form (3) with the provisoes P as one of the premisses, it must be borne in mind that the language and the specific form in which P is expressed are left quite vague. The notation is not meant to be a sharp explication, but rather a convenient way of referring to the subject at issue in the context of an attempt to shed some further light on it.

Note that the proviso P does not include clauses to the effect that the establishment of S^1 has not been affected by errors of observation or measurement, by deceit or the like; that is already implied by the premise S^1 itself, which trivially asserts that S^1 is true. The proviso is to the effect, not that S^1 is true but that it states the *whole* truth about the relevant circumstances present.

Note further that the perplexities of the reliance on provisoes cannot be avoided by adopting a structuralist – or nonstatement – conception of theories broadly in the manner of Sneed and Stegmüller.[11] That conception construes theories not as classes of statements but as deductively organized systems of statement functions, which make no assertions and have no truth values. But such systems are presented as having empirical models – for example, the solar system might be claimed to be a model of a structuralist formalization of Newtonian celestial mechanics. But a formulation of this claim – and its inferential application to particular astronomical occurrences – again clearly assumes the fulfillment of pertinent provisoes.

8. METHODOLOGICAL ASPECTS OF PROVISOES

The elusive character of proviso clauses raises the question of how a theoretical inference of type (3) can be applied to particular occurrences – and, more specifically, on what grounds a proviso P may be taken to be satisfied or violated in specific cases.

There are circumstances that provide such grounds. If the theory T has strong previous support but if its application to a

[11] Compare Sneed (1979), Stegmüller (1976), especially chap. 7.

new case yields incorrect predictions S^2, then doubts may arise about S^1; but in the absence of specific grounds for such doubts, a violation of P (i.e., the presence of disturbing factors) may suggest itself. If this conjecture can be expressed in the language of the theory T and if replacing S^1 by a correspondingly modified sentence $S^{1'}$ yields successful predictions, then this success will constitute grounds for attributing the predictive failure of the original theoretical inference to a violation of its proviso clause.

Thus, the failure of Newton's otherwise highly successful theory to predict certain perturbations in the orbit of Uranus in terms of the gravitational attraction exerted on it by the sun and by the planets known before 1846 led to the conjecture of a proviso-violation – namely, the assumption that Uranus was subject to the additional attraction of a hitherto unknown planet, a conjecture borne out by the subsequent discovery of Neptune.

Sometimes predictive failure of a theory is attributed to proviso-violations even though the presumably disturbing factors cannot be adequately specified.

Consider, for example, the controversy between Robert A. Millikan and Felix Ehrenhaft over the results of the famous experiments in which Millikan measured the rates at which small electrically charged oil drops rose and sank in the electric field between two horizontal electrically charged metal plates. From those rates he computed, by means of accepted theoretical principles, the size of the charges of those oil drops and found that all of them were integral multiples of a certain minimum charge e, whose numerical value he specified. Millikan presented his findings as evidence for the claim that electricity had an atomistic structure and that the atoms of electricity all had the specified charge e.

Ehrenhaft objected that in similar experiments, he had found individual charges which were not integral multiples of Millikan's value e and which, in fact, were often considerably smaller than e, suggesting the existence of "sub-electrons."[12]

---

[12] Millikan gives a detailed account of his investigations in Millikan (1917), in which Ehrenhaft's claims are discussed in Chapter VIII. The controversy is examined in a broader scientific and historical perspective in Holton (1978).

Ehrenhaft accordingly rejected Millikan's theoretical claims T on grounds of predictive failure.

Millikan replied in careful detail. Referring to difficulties he had encountered in his own work, he argued that Ehrenhaft's deviant results could be due to disturbing factors of various kinds. Among them, he mentions the possibility that tiny dust particles might have settled on the falling oil droplets, thus changing the total force acting on them; the possibility that evaporation might have reduced the mass of an observed drop; the possibility that the strength of the electric field might have decreased as a result of battery fatigue; and so forth.

Ehrenhaft repeated his experiments, taking great pains to screen out such disturbing factors, but he continued to obtain deviant findings. The sources of these deviations have never been fully determined; in fact, Ehrenhaft's results turned out not to be generally reproducible. Millikan's ideas, on the other hand, were sustained in various quite different applications. Thus, eventually Ehrenhaft's claims were gradually disregarded by investigators in the field, and Millikan won the day and the Nobel Prize.

Interestingly, as has been pointed out by Holton,[13] Millikan himself had recorded in his laboratory diaries several sets of quite deviant measurements, but he had not published them, attributing them to disturbing factors of various kinds and sometimes not even offering a guess as to the source of the deviation.

But evidently, it cannot be made a *general* policy of scientific research to attribute predictive failures of theoretical inferences to the violation of some unspecified proviso; for this "conventionalist strategem," as Popper has called it, would deprive a theory of any predictive or explanatory force.[14]

I think that at least in periods of what Kuhn calls normal science, a search for disturbing influences will consider only factors of such kinds as are countenanced by one or another of the currently accepted scientific theories as being nomically relevant to the phenomena under consideration.

[13] Holton (1978, esp. pp. 58–83).
[14] See, for example, Popper (1962, pp. 33–9); Stegmüller (1976, chap. 14).

Thus, if a prediction based on Newtonian mechanics fails, one might look for disturbing gravitational, electric, magnetic, and frictional forces and for still some other kinds, but not for tele-kinetic or diabolic ones. Indeed, since there are no currently accepted theories for such forces, we would be unable to tell under what conditions and in what manner they act; consequently, there is no way of checking on their presence or absence in any particular case.

The mode of procedure just mentioned is clearly followed also in experiments that require the screening-out of disturbing outside influences – for example, in experimental studies of the frequency with which a certain kind of subatomic event occurs under specified conditions. What outside influences – such as cosmic rays – would affect the frequency in question and what shielding devices can serve to block them (and thus to ensure satisfaction of the relevant proviso) is usually determined in the light of available scientific knowledge, which again would indicate no way of screening out, say, telekinetic influences.

If a theory fails to yield correct predictions for a repeatable phenomenon by reference to factors it qualifies as relevant, then certain changes within that theory may be tried, introducing a new kind of nomically relevant factor. Roentgen's discovery of a photographic plate that had been blackened while lying in a closed desk drawer is, I think, a case in point; it led to the acknowledgment of a new kind of radiation.

Finally, persistent serious failures of a theory may lead to a revolution in Kuhn's sense, a revolution which places the phenomena into a novel theoretical framework rather than modifies the old one by piecemeal changes. In this case, the failures of the earlier theory are not attributed to proviso violations; indeed, it is quite unclear what such an attribution would amount to.

Consider a theoretical inference that might have been offered some 250 years ago on the basis of the caloric fluid theory of heat or the phlogiston theory of combustion. The relevant provisoes would then have to assert, for example, that apart from the factors explicitly taken into account in the inference, no other factors are present that affect, say, the flow of caloric fluid between bodies or the degree of dephlogistication of a body. But from our present vantage point, we have to say that there are no such substances as caloric fluid or phlogiston – and that, therefore, there could be no proper proviso claim of the requisite sort at all.

And yet, it appears that the claims and the inferential applications of any theory have to be understood as subject to those elusive provisoes.

There is a distinct affinity, I think, between the perplexing questions concerning the appraisal of provisoes in the application of scientific theories and the recently much discussed problems of theory choice in science.

As Kuhn in particular has argued in detail, the choice between competing theories is influenced by considerations concerning the strength and the relative importance of various desirable features exhibited by the rival theories; but these considerations resist adequate expression in the form of precise explicit criteria. The choice between theories in light of those considerations, which are broadly shared within the scientific community, is not subject to – nor learned by means of – unambiguous rules. Scientists acquire the ability to make such choices in the course of their professional training and careers, somewhat in the manner in which we acquire the use of our language largely without benefit of explicit rules, by interaction with competent speakers.

Just as, in the context of theory choice, the relevant idea of the superiority of one theory over another has no precise explication even though its use is strongly affected by considerations shared by scientific investigators, so too in the inferential application of theories to empirical contexts, the idea of the relevant provisoes has no precise explication; yet this idea of the relevant provisos is by no means arbitrary, and its use appears to be significantly affected by considerations akin to those affecting theory choice.

REFERENCES

Carnap, Rudolf
  1950: *Logical Foundation of Probability* (Chicago: University of
    Chicago Press).

1956: "The Methodological Character of Theoretical Concepts," in H. Feigl and M. Scriven, eds., *Minnesota Studies in the Philosophy of Science*, vol. 1 (Minneapolis: University of Minnesota Press), pp. 38–76.

1963: "Arthur Pap on Dispositions," in Paul A. Schlipp, ed., *The Philosophy of Rudolf Carnap* (La Salle, IL: Open Court), pp. 947–52.

1966: *Philosophical Foundations of Physics* (New York: Basic Books).

Craig, W.

1956: "Replacement of Auxiliary Expressions," *Philosophical Review* 65, 38–55.

Fiegl, H.

1970: "The 'Orthodox' View of Theories: Remarks in Defense as Well as Critique," in M. Radner and S. Winokur, eds., *Minnesota Studies in the Philosophy of Science*, vol. 4 (Minneapolis: University of Minnesota Press), pp. 3–16.

Hempel, Carl G.

1958: "The Theoretician's Dilemma," in H. Fiegl, M. Scriven, and G. Maxwell, eds., *Minnesota Studies in the Philosophy of Science*, vol. 2 (Minneapolis: University of Minnesota Press), pp. 37–98; reprinted in C. G. Hempel, *Aspects of Scientific Explanation* (New York: The Free Press, 1965), pp. 173–226.

1969: "On the Structure of Scientific Theories," in *Isenberg Memorial Lecture Series 1965–66* (East Lansing: Michigan State University Press), pp. 11–38.

1970: "On the 'Standard Conception' of Scientific Theories," in M. Radner and S. Winokur, eds., *Minnesota Studies in the Philosophy of Science*, vol. 4 (Minneapolis: University of Minnesota Press), pp. 142–63.

Holton, Gerald

1978: "Subelectrons, Presuppositions, and the Millikan-Ehrenhaft Dispute," in G. Holton, *The Scientific Imagination* (Cambridge, UK: Cambridge University Press), pp. 25–83.

Millikan, Robert A.

1917: *The Electron* (Chicago: University of Chicago Press, 1963, facsimile edition).

Pap, A.

1963: Reduction Sentences and Disposition Concepts," in Paul

A. Schlipp, ed., *The Philosophy of Rudolf Carnap* (La Salle, IL: Open Court), pp. 559–97.

Popper, Karl R.

1962: *Conjectures and Refutations* (New York/London: Basic Books).

Putnam, H.

1962: "What Theories Are Not," in E. Nagel, P. Suppes, and A. Tarski, eds., *Logic, Methodology and Philosophy of Science* (Stanford: Stanford University Press), pp. 240–51.

1965: "Craig's Theorem," *The Journal of Philosophy* 62, 251–60.

Ramsey, F. P.

1931: R. B. Braithwaite, ed., *The Foundations of Mathematics and Other Logical Essays* (London: Routledge and Kegan Paul, 1931; Paterson, NJ: Littlefield, Adams, paperback reprint 1960), pp. 212–36. Also in D. H. Mellor, ed., *Philosophical Papers* (Cambridge, UK: Cambridge University Press, 1990), pp. 112–36.

Sneed, J. D.

1979: *The Logical Structure of Mathematical Physics*, 2nd ed. (Dordrecht: D. Reidel).

Stegmüller, W.

1976: *The Structure and Dynamics of Theories* (New York: Springer Verlag).

Suppe, F.

1974: "The Search for Philosophic Understanding of Scientific Theories," in F. Suppe, ed., *The Structure of Scientific Theories* (Urbana: University of Illinois Press); 2nd Edition, with an added "Afterword" (1977).

Memoirs

13 [70]* Rudolf Carnap, "Logical Empiricist," *Synthese* 25 (1973), 256–68.

14 [89] "Der Wiener Kreis und die Metamorphosen seines Empirismus," in Norbert Leser, ed., *Das geistige Leben Wiens in der Zwischenkriegszeit* (Wien: Bundesverlag, 1981), pp. 205–15. Translated by Christian Piller.

15 [106] "Hans Reichenbach Remembered," *Erkenntnis* 35 (1991), 5–10.

16 [109] "Empiricism in the Vienna Circle and in the Berlin Society for Scientific Philosophy: Recollections and Reflections," *Scientific Philosophy: Origins and Developments*, in Friedrich Stadler, ed., Kluwer (Dordrecht: Academic Publishers, 1993), pp. 1–9.

In the end, Carnap and Neurath had more influence than his *Doktorvater*, Reichrenbach, on Hempel's thought; and although the interaction with Carnap went on much longer, it was Neurath whose influence proved to be the more enduring. Hempel tells the story here in memorial essays (13, 15) on Carnap and Reichenbach – and in two historical essays (14, 16) in which Neurath figures prominently on the career of logical empiricism.

---

* The numbers in brackets represent the publication number for each chapter in this list. (See the section entitled "C. G. Hempel's Publications" beginning on p. 305.)

Chapter 13
Rudolf Carnap
*Logical Empiricist**

Rudolf Carnap was the leading figure among the originators and the moving spirits of the stream of philosophical thought known as logical positivism or logical empiricism. Carnap preferred the latter name: The appellation "empiricism" rather than "positivism" was to set the movement apart from earlier forms of positivism, and the term "logical" was to call attention to the great importance this new empiricism attributed to the concepts and methods of contemporary logic as tools of philosophical analysis.

In Carnap's work, logic loomed large not only as a tool but also as a subject of philosophical investigation. A great deal of his research was devoted to problems in logic and metalogic, including the fields of logical syntax and semantics; the results of this work comprise substantial contributions to deductive logic as well as the most comprehensive and rigorous system of inductive logic yet devised.

The papers of my colleagues on this symposium address themselves to Carnap the logician; let me, therefore, attempt a brief appreciation of Carnap the logical empiricist philosopher.

Carnap's work outside the field of logic was devoted almost exclusively to epistemology and the philosophy of science, and his principal contributions to these fields are united by a common

* From Carl G. Hempel, "Rudolf Carnap, Logical Empiricist," *Synthese* 25 (1973), 256–68. Reprinted with the kind permission of Kluwer Academic Publishers.

leitmotiv – namely, the search for ever more careful and philosophically illuminating reformulations, or explications, of the basic idea of empiricism that all our knowledge of the world ultimately derives from what is immediately given to us in the data of our direct experience.

Stated in these general terms, the idea could be construed as a psychologic-genetic claim concerning the development of man's conception of the world; but Carnap characteristically presented empiricism as a systematic-logical claim to the effect that all the concepts suited to describe the world – and thus, all the concepts that could ever be required by empirical science, from physics to sociology and historiography – can be reduced, in a clearly specifiable sense, to concepts serving to describe the data of immediate experience or observation; and that analogously all our beliefs or assertions about the world can be reduced, in clearly specifiable ways, to beliefs or assertions concerning the data of immediate experience.

This wording is still rather vague. Carnap's standards of clarity and rigor would require that, at least ideally, an explication of the empiricist thesis should meet the following conditions: First, it should characterize a language, with clearly specified logical structure and semantical interpretation, in which both the findings of direct experience and the statements of empirical science can be expressed; second, it should delimit, within that language, a basic experiential vocabulary by means of which the data of immediate experience or observation can be described; and third, it should give a precise logical characterization of the senses in which the terms and the statements of empirical science are claimed to be reducible to experiential terms and to experiential statements, respectively.

Any specific characterization of the two kinds of reducibility will clearly yield a necessary condition for a corresponding construal of "empirical significance" for terms and for sentences – a concept which played a central role in the logical empiricists' early critique of speculative metaphysics and which continued to inform their persistent attempts to characterize scientific knowledge so as to ensure empirical grounding and bearing for scientific theories at a minimal cost in logical constraints.

The problem of explicating the relationship between empirical knowledge and the immediate data of our experience has thus been given a linguistic turn, which is characteristic of Carnap's general conception of the nature of philosophical issues. The problem now concerns, broadly speaking, certain logical and semantic features of languages suitable for the description of all empirical phenomena – and thus, of languages suitable for the purposes of empirical science.

In his first major work, *Der logische Aufbau der Welt*,[1] Carnap chose, from among various possibilities he considered acceptable, a phenomenalistic basis: He assumed the data of direct experience to be described by a phenomenalistic vocabulary, and he argued that all terms required for the description of empirical phenomena can be explicitly defined on this basis. In fact, in a striking tour de force, he limited his primitive vocabulary to one single predicate, one standing for the binary relation of remembered partial similarity between what he called elementary experiences. Reducibility of terms was thus claimed in the very strict sense of explicit definability. This definability thesis for terms immediately implies that all empirical statements are translatable into sentences containing only the phenomenalistic terms chosen as basic.

The *Aufbau* was inspired especially by the construction of mathematics from logic in Frege's work and in *Principia Mathematica* and by Russell's ideas on the definitional reconstruction of the physical world on a phenomenalistic basis. What made Carnap's book such an outstanding achievement – a "dazzling sequel to *Our Knowledge of the External World*," as Quine has called it[2] – is that Carnap did not just offer a general program for a definitional "construction of the world," but actually formulated explicit definitions for a considerable array of concepts; these belong to the lowest domain of the constructional hierar-

---

[1] Rudolf Carnap, *Der logische Aufbau der Welt* (Berlin-Schlachtensee; Weltkreis, 1928; Hamburg: Felix Meiner, 1961); Rolf A. George, trans., *The Logical Structure of the World and Pseudoproblems in Philosophy* (Berkeley: University of California Press, 1967).

[2] W. V. Quine, "Russell's Ontological Development," *The Journal of Philosophy* 63 (1966), 657–67; quotation from p. 667.

chy (i.e., the realm of "autopsychological objects"). The definitional steps intended to lead from this domain of subjective experience to the public spatio-temporal world of physics, to other persons and other minds, and on to socio-cultural phenomena were characterized in a more sketchy manner.

It has since become clear – not least as a result of Carnap's own subsequent research – that the basic theses of the *Aufbau* are too strong to be tenable; and for the recent English edition of his work,[3] Carnap wrote a special preface in which, with characteristic candor, he pointed out what he considered to be their principal shortcomings.

One of these concerned the thesis of definability. As early as 1935, Carnap pointed out that within a language with an extensional logic, dispositional properties are not explicitly definable in terms of manifest characteristics that are symptomatic of them, and he argued that dispositional terms could be introduced into the language of science by means of so-called reduction sentences, which he regarded as affording partial or imcomplete definitions. In his classical essay, "Testability and Meaning,"[4] Carnap developed this idea systematically and in detail. His earlier thesis of the definability of all empirical terms was replaced by the broader one that all empirical terms can be introduced by chains of reduction sentences on the basis of an observational vocabulary, as well as that the translatability claim for the sentences of empirical science gave way to a correspondingly weaker one, roughly to the effect that every empirical statement is at least partially confirmable by evidence sentences of molecular form containing only observational terms. The earlier construals of empirical significance in terms of verifiability or falsifiability were modified accordingly.

As for the experiential basis of empirical knowledge, a phenomenalistic construal had initially recommended itself to

---

[3] Carnap, *The Logical Structure of the World*, pp. viii–ix. Further reflections by Carnap on the *Aufbau* may be found in Paul A. Schlipp, ed., *The Philosophy of Rudolf Carnap* (La Salle, IL: Open Court, 1963), pp. 16–20; and in Carnap's response (pp. 944–7) to Goodman's essay, "The Significance of *Der logische Aufbau der Welt*," also in *The Philosophy of Rudolf Carnap*, pp. 548–58.

[4] In *Philosophy of Science* 3 (1936), 419–71 and 4 (1937), 1–40.

Carnap for reasons that included the apparent promise of providing empirical science with a secure bedrock foundation consisting of phenomenalistic sentences established by immediate experience – and thus indubitably true.[5] But Carnap soon came to share the view held by Neurath and by Popper that the idea – or the ideal – of such empirical certainty is illusory, that every empirical statement is open, in principle, to revision; and thenceforward he preferred to construe the basic evidence for empirical knowledge as expressed in a "thing language" by means of sentences in which directly but publicly observable attributes are ascribed to publicly observable physical objects.

The resulting explication of empiricism thus took the form of physicalism, according to which the thing language, or more broadly the language of physics, affords a unitary language for all empirical science in the sense that all scientific terms can be introduced on the basis of a physical vocabulary by means of definitions or – in a weakened version of the thesis – by means of reduction sentences; and that all scientific statements are translatable into, or – in the weaker version – at least confirmable by, sentences containing only physical terms. For example, psychological terms like *angry* or *excited* were said to refer to bodily states characterized by the disposition to show specific kinds of physical behavior under specific physical conditions. Carnap regarded it as a long-range aim of scientific research, however, to replace such dispositional characterizations by microstructural ones, which would again be formulated in physical terms.[6]

More recently,[7] Carnap pointed out that the construal thus

---

[5] Compare Carnap in Schlipp, op. cit., p. 57.

[6] See Carnap, "Psychologie in physikalischer Sprache," *Erkenntnis* 3 (1932–33), 107–42; an English translation, with a postscript added in 1957, appeared in A. J. Ayer, ed., *Logical Positivism* (New York: The Free Press, 1959), pp. 165–98.

This article affords an excellent example of Carnap's lucid and systematic manner of presenting, illustrating, and supporting a philosophical thesis, of defending it against objections, and of pointing out its implications for certain special topics – such as the character of introspective psychology and the "physicalization of graphology."

[7] See Carnap, "The Methodological Character of Theoretical Concepts," in H. Feigl and M. Scriven, eds., *Minnesota Studies in the Philosophy of Science*, vol. 1 (Minneapolis: University of Minnesota Press, 1956), pp. 38–76.

arrived at does not do justice to the logical character of the terms and sentences of scientific theories, and he, therefore, proposed a further liberalization. He characterized the introduction of theoretical terms as being effected by specifying (I) a set of theoretical axioms containing the basic theoretical terms as primitives, and (II) a set of "rules of correspondence" which afford an empirical interpretation by linking some of the theoretical terms to observational ones.[8] These interpretative sentences, however, need not consist in chains of reduction sentences; indeed, Carnap imposed no explicit general restrictions on their form. In accordance with this conception, Carnap formulated new criteria of empirical significance for theoretical terms and sentences; these criteria are multiply relativized, making the empirical significance of theoretical expressions depend, among other things, on the axioms and the correspondence rules of the theory in which they function.[9] This conception, reflecting a deeper exploration of the logic of scientific theorizing, is a far cry from the simple original construal of empirical significance as a property of certain terms and sentences.

In his autobiography, Carnap remarks that even these newer criteria of significance "are not yet in final form"; but he reaffirms his confidence in the possibility of further clarification. "I am inclined to believe," he says, "that it is somehow possible, even in the wider framework of the theoretical language, to make a clear distinction between those terms and sentences which are cognitively significant and those which are not."[10]

Carnap similarly retained his confidence in the tenability and, indeed, in the philosophical importance of another idea which had played a prominent role since the initial stages of logical

---

[8] Carnap noted later that the methods he had suggested in the *Aufbau* for the ascent from the subjective-phenomenalistic realm to the intersubjective realm of physical concepts could not always yield explicit definitions and were, in fact, more closely akin to the introduction of theoretical concepts through postulates and correspondence rules. (Cf. Carnap in Schlipp, op. cit., p. 19, and in the preface to *The Logical Structure of the World*, pp. viii–ix.)

[9] See Carnap, "The Methodological Character of Theoretical Concepts," sections VI, VII, VIII.

[10] In Schlipp, op. cit., pp. 80–1.

HE CONTINUES TO DEFEND AN-SYN DISTINCTION

empiricism and which was often held to mark an advance over earlier forms of empiricism, such as that of John Stuart Mill. This is the idea that all cognitively significant sentences can be sharply divided into those which have empirical content (and thus are synthetic) and those which are analytically true or false, the analytic truths including as a proper subclass the truths of logic and of mathematics. Again and again, at various stages in the evolution of his ideas, Carnap returned to the task of justifying and explicating the analytic-synthetic distinction; he even proposed an ingenious way of dividing a scientific theory, with the help of its Ramsey-sentence, into an analytic and a synthetic component. The importance this problem possessed for Carnap and the intensity of the effort he devoted to its clarification can be sensed from the words with which he introduced his proposed solution:

> During my work on the article ["The Methodological Character of Theoretical Concepts"] and subsequently, I long searched in vain for a solution to this problem; more specifically, for a general method for analyzing the total postulate set [consisting of the theoretical axioms and the correspondence rules] into two components: analytic meaning postulates . . . for the theoretical terms, and synthetic *P-postulates* . . . which represent the factual content of the theory . . . I believe now to have found a solution for this problem.[11]

The analytic-synthetic distinction and the notion of empirical significance, as is well known, have been subjects of considerable philosophical controversy for the past several decades. Carnap was an active participant in the debate and sought to meet doubts and objections by suitably modifying his explications, while maintaining his conviction that the *explicanda* were reasonably

[11] Schlipp, op. cit., p. 964. The proposed method of analysis is set forth in more detail in Chapter 28 of Carnap, *The Philosophical Foundations of Physics*, Martin Gardner, ed. (New York and London: Basic Books, 1966). At the end of that chapter, too, Carnap speaks of his "many years of searching" for a satisfactory answer, and he concludes: "No difficulties have yet been discovered in this approach. I am now confident that there is a solution and that, if difficulties appear, it will be possible to overcome them" (pp. 272–3).

clear and of great philosophical importance. The debate over this issue is continuing; but no matter what its outcome may be, there is no doubt that Carnap's efforts to clarify those ideas – and more generally his search for ever more adequate construals of empiricism – have considerably deepened and refined our appreciation of the logical complexity of science and of the connections between its branches.

Carnap's method of philosophical analysis might be characterized as explication by logical reconstruction in model languages with precisely specified syntactical and semantic features. The standards he set himself in the use of this method reflect the characteristically clear, precise, and systematic manner in which he sought to express, develop, and defend his ideas on *any* subject. In writing as in teaching and in oral discussion, Carnap evinced a deep faith in the power of reason and of rational critical argument. It was this search for maximal clarity that made the reliance on formal logic and artificial model languages attractive – and, indeed, often indispensable – to him in tackling philosophical issues.

There could hardly have been a more striking contrast in philosophical style within the same school of thought than that between Carnap and Otto Neurath, an influential figure in the earlier stages of logical empiricism, whose formulations were suggestive but vague; whose argumentation, though provocative and often persuasive, tended to be loose, sketchy, and programmatic; who looked at Carnap's reliance on precise model languages with respect but deep misgivings; who spoke of the language of empirical science as a "universal slang" not governed by precise syntactical and semantical rules; and whose metaphor of a ship that is forever being reconstructed on the high seas without benefit of a drydock[12] reflected his conviction that we cannot start building a scientific language from scratch and with Carnapian precision because of the ever-present fuzziness of the ordinary language in which we have to describe our con-

---

[12] The metaphor occurs in Neurath's article "Protokollsaetze," *Erkenntnis* 3 (1932–3), 204–14; English translation in Ayer, *Logical Positivism*, pp. 199–208.

struction;[13] and that, furthermore, scientific knowledge does not rest on a solid foundation of unquestionable truths – a view with which, as noted earlier, Carnap came to agree. It is characteristic of Carnap's undogmatic openmindedness that despite the striking difference in intellectual style, he appreciated Neurath's often shrewd and perceptive judgment, was on cordial personal terms with him, and repeatedly acknowledged his indebtedness to him in his writings.

Carnap was always conscientious and indeed generous in acknowledging the ideas of others that had influenced his thinking; on the other hand, I do not know of a single instance in which he complained about not having received due credit for his ideas or for their priority. About his punctiliousness in the matter, there is a story that in a course he gave at Harvard, Carnap said at one point, "Let *A* be some physical body, such as a stone, or a tree, or – to take Russell's example – a dog."

In his commitment to careful formulation and rigorous reasoning, in his insistence that philosophical theses should be stated and argued for in ways permitting of objective critical appraisal, Carnap evinced an attitude that is often said to be the mark of a good scientist. Indeed, Carnap like other logical empiricists – several of whom had had extensive scientific training – called for a scientific approach to philosophy, and in the days of the Vienna Circle, members of the group often used the appellation *Wissenschaftliche Weltauffassung* to characterize their conception of philosophy and philosophizing.

[13] In a similar vein, Neurath stressed that scientific procedure "does not depend upon 'exactness' but only upon the permanence of scientific criticism. New ideas of scientific importance start mostly with vague and sometimes queer explanations; they become clearer and clearer, but the theories which follow will stand in time before the door with all their new vagueness and queerness." ("Unified Science as Encyclopedic Integration in *International Encyclopedia of Unified Science*, vol. 1, no. 1 [Chicago: University of Chicago Press, 1938], pp. 1–27; quotation from p. 21. This essay, with its markedly pragmatic orientation, reflecting Neurath's historical and sociological interests, contrasts instructively with Carnap's contribution to the same booklet, which is entitled "Logical Foundations of the Unity of Science" [pp. 42–62] and which takes a purely logical and systematic approach to its subject.)

Carnap's predilection for the use of precise model languages in philosophical explication was clearly akin to the keen interest he took in auxiliary international languages, like Esperanto, which in his earlier years he spoke fluently. These international languages, too, were meant to be governed by precisely stated general grammatical and semantic rules without all the usual exceptions, and they were also meant to be simple enough to be readily learned and used. Carnap felt strongly that the wide adoption of such a language would be, among other things, a great help in furthering the international propagation of scientific ideas. In the preface of his *Introduction to Semantics*, for example, he expresses great regret that Tarski's fundamental work on the semantical concept of truth, published in Poland in 1933, had remained virtually unknown outside that country until 1936, because it had been available in Polish only; and he adds, "This fact, incidentally, confirms once more the urgent need for an international auxiliary language, especially for scientific purposes."[14] In speaking with Carnap about the subject, one had the impression that he had examined virtually all of the auxiliary languages that have been proposed since Esperanto, and it was fascinating to hear him compare their strengths and weaknesses.[15] In a conversation about one such language – I believe it was Lancelot Hogben's Interglossa – Carnap once told me that after examining it, he had written the inventor, pointing out what he considered its more important merits and defects; then he added, as an afterthought, that he had written his letter in that language to get some practice.

With that large branch of analytic philosophy which concentrates its efforts on the analysis of ordinary language, Carnap shared the view that various philosophical problems have their roots in peculiarities of ordinary language and in confusions about its use; his critique of speculative metaphysics illustrates that point. In fact, Carnap held that in philosophical explication,

[14] Rudolf Carnap, *Introduction to Semantics* (Cambridge, MA: Harvard University Press, 1942), p. vi.

[15] Section 11 (pp. 67–71) of Carnap's autobiography [in Schlipp, op. cit.], which deals with the subject of language planning, reflects very clearly the various grounds for his interest in international auxiliary languages.

the use of symbolic logic and of constructed languages with explicit syntactical and semantical rules, though "the most elaborate and efficient method, . . . is advisable only in special cases, but not generally," and he considered "the naturalist and the constructionist methods [as] not necessarily competitive, but rather mutually complementary."[16] This point is well illustrated, I think, by the confluence of ideas shown by Carnap's physical-istic-behavioristic analysis of psychological concepts, mentioned above, and by the behavioristic approach taken by Ryle in *The Concept of Mind*.[17] Both proposed dispositional construals of certain psychological terms; and Ryle's criticism of "the ghost in the machine" is very similar in tenor to Carnap's earlier criticism of the idea that behind the physical structure corresponding to such psychological states as excitement, there stands an occult property or power – namely, excitement or the consciousness of it – which itself remains unknowable.[18] Carnap describes this idea as involving "a remarkable duplication . . . : besides or behind a state of affairs whose existence is empirically deter-minable, another, *parallel* entity is assumed, whose existence is not determinable";[19] and he rejects it after critical analysis.

But much of Carnap's work was devoted to philosophical issues concerning disciplines such as mathematics and physics, which have highly developed technical languages of their own; and it seems clear that these are largely beyond the reach of ordinary-language analysis. Moreover, Carnap saw philosophy as aiming at general systematic elucidation rather than, say, at more casuistic efforts to help individual flies out of the particu-lar philosophical fly bottles in which they are trapped.

Against the use of Carnap's logico-constructionist approach to philosophical problems – especially those not concerning techni-

---

[16] Carnap in Schlipp, op. cit., p. 936 and p. 940. These statements occur in Carnap's response to P. F. Strawson's article, "Carnap's Views on Constructed Systems versus Natural Languages in Analytic Philosophy" (in Schlipp, op. cit., pp. 503–18).

[17] Gilbert Ryle, *The Concept of Mind* (London: Hutchinson's University Library, 1949).

[18] See "Psychology in Physical Language," in Ayer, op. cit., p. 173.

[19] Ibid., p. 174.

cal scientific disciplines – it has been argued that ordinary natural languages are too rich, complex, and flexible to fit the rigid Procrustean bed of some formalized linguistic frame. This objection certainly has a point, but it is not quite as telling as it may seem.

1. First, Carnap was far from insisting on just one model of philosophical bed – and that, a Procrustean one. He explicitly countenanced – and called for – the construction and use of philosophical beds of different kinds and degrees of commodiousness. That was the point of "his principle of tolerance."[20]

2. Undeniably, the use of any precise type of model language – to represent, say, the language of physics – involves considerable schematization and simplification; but this is true, to some extent, also of any system of general rules purporting to characterize grammatical features of a natural language.

3. Even if philosophy were to limit itself, casuistically, to helping individual flies out of their particular fly bottles, that philosophical activity or therapy would still have to be informed by general principles. A fly trapped in a bottle or a man trapped in a maze might be led out with his eyes bandaged: He would follow his leader blindly and would eventually find himself outside, but he would not understand how he had been trapped nor how he had eventually got out. But there is no analogue to this mode of physical liberation in the case of a person philosophically trapped in a linguistic maze. The only way to lead him out is with his eyes open, as it were. He has to be *shown* the way out, to use Wittgenstein's phrase (i.e., he must come to understand both what features of the trap led to his getting caught in the first place and how to avoid the same fate in similar situations). And this always requires insights of a *general* type, con-

[20] Carnap first pronounced this principle in *Der logische Syntax der Sprache* (Vienna: Julius Springer, 1934). This book, in a translated version that was extensively revised by Olaf Helmer, was published in English as Amethe Smeaton, Countess von Zeppelin, trans., *The Logical Syntax of Language* (London: Kegan Paul, 1937). Carnap reaffirmed this principle, in various forms, in later writings. The important role it played in his thinking is reflected in the repeated references to it in his "Intellectual Autobiography," in Schlipp, op. cit., pp. 18, 55f, 66, 70.

cerning, for example, linguistic contexts of a certain kind, or language games of a certain sort whose general rules are then projected onto the particular case at hand.

4. As noted above, the precisely characterized languages by reference to which certain philosophical problems have been studied are often distinctly simpler than those required for the purposes of science. For example, Carnap's theory of reduction and confirmability and his vast system of inductive logic are limited to languages with first-order logic, which certainly does not suffice for the formulation of contemporary physical theories. The same remark applies to various studies by Carnap and other empiricists that deal with the structure and function of scientific theories, with the qualitative concept of confirmation, with the logic of scientific explanation, and so forth.

But Carnap often stressed that these studies are intended only as the first stage in the development of more comprehensive theories and that the solutions they offer may well permit of extension to more complex situations. I have heard him remark that Euclid might similarly have encountered the objection that the basic concepts and principles of his geometrical theory were much too simple to permit an adequate account of the infinitely rich diversity of geometrical shapes exemplified in nature; and, indeed, the application of geometry to such complex configurations had to await the development of analytic geometry and of calculus. But these powerful extensions did not show that Euclid's theory had been futile in its simplicity – on the contrary, that theory had been the essential first step.

5. Next, an analysis of a subject by reference to a relatively simple kind of language may reveal unexpected and quite fundamental difficulties. This is illustrated by the problems that Carnap encountered in attempting to define dispositional and theoretical concepts; by the paradoxes of confirmation; by Goodman's riddle; by various problems concerning inductive inference and probabilistic explanation, which arise even for languages of rather simple kinds.

6. Finally, it can be of immense philosophical interest to find that what seems to be a hopelessly difficult problem can, in fact, be solved for languages of certain precisely characterized kinds.

One classical example is Tarski's method of constructing, for languages with sharply specified characteristics, a concept of truth that meets his condition of adequacy and escapes the semantic paradoxes. Another striking illustration is provided by one of the tasks Carnap tackled in his work on inductive logic – namely, to give an explicit definition of inductive probability which assigns to any sentence h in a given language a definite numerical probability $p(h,e)$ with respect to any noncontradictory 'evidence-sentence' e in the same language – and in such a way, moreover, that the fuction p thus defined has the mathematical properties of a conditional probability. The demonstration that this can be done for languages of the kind studied by Carnap is surely a result of great philosophical significance.

Let me conclude with a few words about Carnap the teacher and the man. I had the good fortune to be among Carnap's students in Vienna, in Prague, and later once more in Chicago. What made me decide, despite difficulties, to go to Vienna at least for a semester was the powerful impact of his *Aufbau* and of his shorter book, *Scheinprobleme in der Philosophie*,[21] both of which had just recently been published. Carnap's courses differed strikingly from those I had taken with some of the most eminent philosophers in Germany. In my efforts to expand into a coherent text the notes I had taken during their lectures, I often had a sense of practicing an intricate ritual dance with words, whose message was elusive or obscure. In Carnap's teaching, all was order and light. His presentation and discussion were systematic, lucid, and incisive; he offered careful arguments for his views and was willing to listen seriously to doubts and objections from his students. In the traditional seminars, the students were often expected to expound the ideas of this or that philosopher, while critical comments on those ideas – not to mention the views of the professor in charge – were not encouraged. The spirit that prevailed in Carnap's seminars – and also in those of some of my

[21] Carnap, *Scheinprobleme in der Philosophie* (Berlin-Schlachtensee: Weltkreis, 1928). The volume mentioned as the first item in note 3 above (i.e., *The Logical Structure of the World*) contains an English translation of this short work as well as of the *Aufbau*.

other teachers, among them Reichenbach, Schlick, and Waismann – was, therefore, wonderfully stimulating and exhilarating.

I have often wished that Carnap had reached a wide audience as a teacher. He tended to think of himself as lacking the skill to present philosophical ideas to beginning students or to a more general interested public. Yet at the German University in Prague, he gave an introductory course in the history of modern philosophy in which I recall hearing some lectures that dealt most interestingly and lucidly with Kant's First *Critique*. But Carnap did feel most at home with advanced students; and in his work with them, his strength as an inspiring teacher shone forth most powerfully.

No one who came to know Carnap even moderately well could fail to sense his unusual human stature. He was a man entirely innocent of pretentiousness, self-importance, or pettiness; and he possessed a genuine and guileless openness to the ideas and problems of others, a man moved by an intense intellectual curiosity and a deep urge for conceptual order and understanding, but equally inspired by abiding social and humanitarian concerns, and unfalteringly faithful to his moral convictions. Carnap was a rare and a great man.

Chapter 14

The Vienna Circle and the
Metamorphoses of Its Empiricism

1. INTRODUCTORY REMARKS

The Vienna Circle was made the topic of this lecture series for good reasons. The philosophical ideas of that group, which started to blossom during the years between the wars and soon gained international influence, belong, without doubt, to the most important and most fruitful manifestations of Vienna's intellectual life of that time.

That I have been given the honorable task of speaking here about the Vienna Circle and about the evolution and influence of its ideas is certainly due in part to the fact that I belong to the nowadays small group of people who personally knew or know the leading thinkers of the Vienna Circle and similar-minded groups.

Furthermore, my intellectual development was strongly influenced by the thinkers of the Vienna Circle, as well as by Hans Reichenbach, the leading personage of the Berlin Group of empiricist philosophers. I feel a deep respect for and gratitude toward those thinkers; but gradually, I came to question their basic ideas, as I have indicated in my writings. But this was entirely in the spirit of my teachers, who always encouraged their students toward independent and critical thinking.

In view of my close connection with the movement of logical positivism or, as it was later called, logical empiricism, I hope it will be seen as appropriate if I start with some personal recollections about the leading representatives of the Vienna Circle.

268

2. SOME PERSONAL MEMORIES

My first contact with the Vienna Circle was stimulated by reading Carnap's book *The Logical Structure of the World*. This work made such a deep impression on me that I decided to interrupt, at least for a while, my studies with Reichenbach and other philosophers, mathematicians, and physicists in Berlin, and to go to Vienna for the winter term (1929/30). I attended lectures and seminars by Schlick, Carnap, and Waismann, and I regularly participated in the meetings of the Vienna Circle.

Among the members who were very active at that time I remember especially Moritz Schlick, Rudolf Carnap, Otto Neurath, the mathematician Hans Hahn, Friedrich Waismann, and my old friend Herbert Feigl, who was then Schlick's assistant and who later, in the early 1930s, moved to the United States, where he became an active and influential proponent of logical positivism. It seems to me that the logician Kurt Gödel, the mathematician Karl Menger, and the social scientist Felix Kaufmann, all of whom later emigrated to the United States, were also sometimes present at those meetings.

Karl Popper, too, was in Vienna then and was engaged in a lively exchange of ideas with the members of that group. But Popper always disassociated himself from the Vienna Circle and from logical positivism – unfortunately in a too vigorous and too polemic manner, I think. Of course, there were some differences of opinion between Popper and the members of the Vienna Circle, but there were also strong affinities. Moreover, as I will point out in some detail later, considerable differences of opinion were also common among members of the Vienna Circle: There was no party doctrine, no list of philosophical theses the members had to swear by.

What was common to these thinkers was a collection of basic ideas, not sharply defined, that inspired and directed their philosophical efforts. Among them was an empiricist conception of knowledge and the conviction that scientific research, especially as done in the exact natural sciences, is the most highly developed form of the search for knowledge. From an intellectual viewpoint, they all noted with strong dissatisfaction that philosophy, especially metaphysics, was having very little success in its attempt to solve the problems that were considered central. Finally, they shared the conviction that within the framework of a precisely formulated empiricism, all these problems could, with the help of precise, analytical methods expressible in mathematical logic, either be solved or shown to be pseudoproblems without genuine subject matter.

Within this setting considerable differences existed within the group – as well as in their style of philosophical argument.

The differences between Neurath and Carnap provide a good example of what I have in mind. Neurath was a tall man, a bit corpulent, but very agile. He was extroverted, lively, and immensely energetic and enterprising. He was always full of plans and ideas, and he played an important role in organizing conferences, in which members of the Vienna Circle, members of the Berlin Group surrounding Reichenbach, and members of the Warsaw School participated – as well as other thinkers, both critical and sympathetic. Neurath was also essentially involved in the planning and the publication of the *International Encyclopedia of Unified Science*, which was begun in Chicago in 1937/38.

Whereas Schlick and Carnap, as well as Reichenbach in Berlin and the theoretical physicist Philipp Frank in Prague (who was also very close to the Vienna Circle), had come to philosophy from mathematics and physics, Neurath had a special education in the social sciences, and he was especially interested in problems of sociology and economics.

Neurath had deep social and political convictions that were inspired by Marxism. Yet he was not a Marxist in a dogmatic sense, as was also pointed out by Carnap in his intellectual autobiography. According to Neurath, every theory had always to be further criticized and put to further test.[1]

Neurath was keenly aware of the possibility of misusing theological and metaphysical ideas for social and political control, and it is the importance he attached to avoiding any such

---

[1] Compare "The Vienna Circle (1926–35)" in Paul A. Schlipp, ed., *The Philosophy of Rudolf Carnap* (La Salle, IL: Open Court, 1963), pp. 20–34.

misuse that may well explain the fact that his philosophical and methodological views contained some extreme and overly simplified ideas.

Inspired by materialism, Neurath characterized psychology and the social sciences in a way sharply opposed to there being any fundamental difference between the natural sciences on one side and the humanities on the other. Neurath thought that all scientific disciplines – from physics to sociology and history – deal with the activities of more or less complex physical systems (among them humans and groups of humans) and that, therefore, all scientific results had to be expressible in a "uniform physicalistic language." He was convinced that the use of mentalistic and other nonphysicalistic expressions always involved the danger of conceptual confusion and of empty and futile metaphysical speculation (e.g., about how an immaterial "spirit" could influence a person's bodily behavior). Neurath had deep respect for Max Weber, but he held that a detailed examination was required to determine which of Weber's ideas could be expressed in a physicalistic language and, thus, really had empirical content.

In consequence, Neurath also rejected many classical formulations of dialectical materialism because they were metaphysical, but he was concerned to reformulate them in an acceptable – that is, a physicalistic – way.

In the same way he regarded many theses of Wittgenstein's *Tractatus Logico-Philosophicus* as purely metaphysical, devoid of content, and thus neither true nor false. During the earlier discussions of the Vienna Circle, which closely examined the *Tractatus*, Neurath objected again and again that it was pure metaphysics they were engaged in. When Schlick told him that his interruptions were disturbing, Neurath offered just to say "M!" when the discussions turned metaphysical. But soon he came up with an improvement: "I think," he said, "it would save us all time if I said 'Non-M!' in case we are not doing metaphysics for once."

Neurath was deeply convinced of the importance of his physicalistic program. He asked his readers to participate in the big and liberating task of translating what psychology and the social sciences say into a physicalistic terminology in order to exhibit their empirical content.

In Neurath's writings, though, the reader will not often find precise formulations and detailed justifications of his physicalism. Neurath developed his ideas – in many ways important and fruitful ideas – in a vivid, interesting, and suggestive style that nevertheless looked sketchy from a systematic perspective. In this respect, his philosophical style resembled that of a man who explains and defends a comprehensive social-political program.

On the other side of the spectrum of philosophical style was Carnap. He was a somewhat introverted person and an extraordinarily thorough, highly systematic, and precise thinker who insisted on formulating philosophical problems as explicitly and clearly as possible as well as on engaging with these problems by means of precise logico-analytical methods. Carnap also held a physicalistically inspired epistemology, but he tried to express it in precise and explicitly justified statements. Because he constantly reworked his ideas, Carnap changed his version of physicalism, as well as his views on other matters, several times, always toward a more liberal position.

For example, around 1930 he developed a version of physicalism according to which mentalistic expressions of psychology, such as "intelligent," "tired," "angry," and "thinking of Vienna," refer to behavioral dispositions that are expressible in physical terms. In this respect these expressions are strictly analogous to physical terms such as "soluble in water," "transparent," and "electrical conductor." The term "soluble in water," for example, refers to an object's disposition to manifest a characteristic physically describable "reaction" (namely, dissolving), given certain physically describable "stimuli" (namely, being put into water). The stimulus to which the disposition "intelligent" refers would be a situation in which some person is confronted with an intelligence test; the reaction characteristic of this disposition would be that this person marks certain of the allowable answers. These stimuli and reactions, it seems, are describable in purely physical terms; the same holds for other psychological terms of the kind indicated by the examples above. The terms in question, therefore, refer to behavioral dispositions describable in purely physical terms and not to immaterial spiritual entities.

In this way, Carnap had already formulated the basic idea of a quasi-behavioristic conception of psychological concepts, one which was later developed in sensitive detail by Gilbert Ryle in his influential book *The Concept of Mind*.[2]

Later, in 1956, Carnap rejected this conception as too narrow. 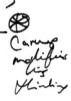 He replaced it with another account, according to which many psychological concepts are – considered from a logical and methodological point of view – of the same nature as the concepts of theoretical physics: They cannot be defined as pure behavioral dispositions, and we have to view them as characterized by theoretical principles.

Carnap was sympathetic to many of Neurath's political and social ideas. But he insisted that all considerations of that sort should be put aside as irrelevant when investigating problems of epistemology and philosophy of science. Carnap insisted that one must justify one's views with the utmost rigor by systematic-logical arguments.

Schlick seemed to me to be a conservative and somewhat reserved personality. But he was never dogmatic or authoritarian, and Carnap attributed the friendly atmosphere that prevailed in the discussions of the Vienna Circle mainly to Schlick's unfailing friendliness, to his tolerance, and to his modesty. In his seminars Schlick was a masterly, sympathetic, and inspiring teacher.

The same is true of Friedrich Waismann, who admirably exemplified the Socratic tradition both in his seminars as well as in personal conversation.

Unfortunately, I have no personal memories of that uncommon thinker who could probably be called the eminence grise of the Vienna Circle – namely, Ludwig Wittgenstein. Much to my regret, I did not see him even once.

The early development of the Vienna Circle was strongly marked by Wittgenstein's ideas; many meetings of the group

[2] Gilbert Ryle, *The Concept of Mind* (London: Hutchinson's University Library, 1949).

were devoted to a thorough and careful sentence-by-sentence discussion of the *Tractatus*. Waismann often played the part of speaker and exegete.

But the philosophy of the Vienna Circle was in no way just Wittgenstein's philosophy. For example, the issue of the verification of empirical sentences was no more mentioned in the *Tractatus* than was the issue of immediate perception or experience, which is the basis of all empirical knowledge for the logical positivists. Furthermore, Carnap and Neurath, for example, rejected Wittgenstein's ideas about the mystical – which is that which cannot be said. Contrary to Wittgenstein, Carnap's description of philosophy as syntactical (later also as semantical) metatheory of the language of science deals with objects of precise and explicitly formulated theories that are, according to the *Tractatus*, inaccessible to language.

The ideas that Wittgenstein developed after 1929, which strongly deviate from his *Tractatus* and were posthumously published beginning in 1953 in his *Philosophical Investigations* and in other works, did not play, as far as I can see, any influential role in the further development of logical positivism. But these ideas had an enduring effect on the analytical philosophy of ordinary language, which, starting in England, spread out from there to other countries within and outside the English-speaking world.

In his intellectual autobiography, Carnap expresses his high esteem for Wittgenstein, and he notes that Wittgenstein's attitude toward theoretical problems was more that of an artist than that of a scientist. When Wittgenstein started to formulate his views on a philosophical problem, Carnap says, one could often see the inner struggle, the painful effort to break through from darkness into light. When Wittgenstein finally stated his answer, it appeared to be a newly created work of art or maybe a divine revelation, and any attempt to give a rational and analytic comment would look like a profanity.

Carnap's extremely thorough and systematic style – to explain almost every concept and to explicitly justify every assertion, even if obvious – looked, without question, unbearably pedantic to Wittgenstein. I think that these great differences in philosophical style were a main reason for Wittgenstein's decision to exclude Carnap from 1929 on from his discussions with Schlick. The final break between these two thinkers occurred, to my knowledge, when Wittgenstein found a book of the parapsychologist Albert von Schreenck-Notzing in Carnap's library. Both Wittgenstein, and interestingly enough, Neurath as well (for other reasons though) thought it unreasonable to deal with such ideas, and they emphasized how silly and trivial are the alleged messages produced in spiritual séances. Carnap agreed on this point, but he thought that the existence and the explanation of parapsychologistic phenomena constitute an important empirical problem, one that could not just be taken off the table because of some intuition or philosophical sensitivity.

Doing philosophy in those years in the milieu of logical empiricism was very exciting, especially for a young student. One felt oneself to be a witness to a big change in philosophy, a change in the promotion of which one actively participated. For students in this group, this feeling was without doubt strengthened by the fact that the leading figures of this philosophical movement, including Reichenbach in Berlin, were always willing to listen to questions and objections and to discuss them seriously. Sometimes one found oneself in a situation where all the participants in a seminar tried to figure out the solution of a problem that the teacher considered important but was not able to find a satisfying answer for on his own. I can say similar things about the attitude of two of my other teachers in Berlin – the psychologists and philosophers of science Wolfgang Köhler and Kurt Lewin, who both shared a great interest in logical empiricism.

The sense of being a participant in an important intellectual movement was further supported by the fact that the ideas of the Vienna Circle aroused great interest in many other thinkers. Philosophers and scientists from England, France, Holland, Scandinavia, Poland, and North America actively participated in the animated and productive conferences organized by the groups in Vienna and Berlin.

Tragically, the strongest impulse for the further spreading of logical empiricism was given by the emigration of members of the groups from Vienna, Berlin, and other places under the terror and ideological madness of National Socialism. The leading thinkers of this movement went mainly to England and America, where their ideas, brought into contact with those of pragmatism, operationalism, behaviorism, and other schools, flourished and found new adherents.

But it was not granted to all who were close to logical empiricism to find a new sphere of activity abroad. I here commemorate my friend Kurt Grelling, a highly talented thinker of the Berlin Group, who perished with his wife in the gas chambers of Auschwitz; and I commemorate the very able young Polish logician Janina Hosiasson-Lindenbaum, who became a victim of National Socialist murderers.

Nazi victims

3. LOGICAL EMPIRICISM: THE MAIN OBJECTIVES

I now turn to a sketchy survey of the central problems and ideas of the Vienna Circle, as well as of the far-reaching changes they gradually underwent.

The central concern of logical empiricism was to provide a clear formulation of and justification for an empiricist theory of knowledge. This theory should do justice to the character of everyday knowledge and especially to technical-scientific knowledge. All the different empiricist studies about the form, content, and testability of scientific hypotheses and theories, about the nature of scientific explanation and understanding in the natural and the social sciences, about induction and probability, about space and time, and about many other subjects can be seen as contributions to a comprehensive empiricist theory of the foundations, the logical structure, and the range and reliability of scientific knowledge.

To put it roughly, the basic claim of empiricism is that all our ideas about the world rest ultimately on our immediate experience. This vague claim can be understood in two ways: as an empirical hypothesis about the genetic and psychological features of scientific thinking, and, on the other hand, as a systematic-logical thesis about the warrant, justification, or critical evaluation of empirical statements. According to logical positivism, the psychological-genetic examination of knowledge is an interesting research task for science, but it has no bearing whatsoever on the philosophy of science and epistemology. In its explicit rejection of all psychological elements – especially clearly expressed by Carnap and Popper – empiricism is a logico-systematic position that can be summarized as follows:

1. All statements about the world (i.e., all empirical statements) are testable by so-called observation sentences or basic sentences that describe the findings of direct observation.

2. All empirical concepts are reducible to elementary observational concepts (i.e., to concepts whose applicability in individual cases can be decided simply on the basis of direct observation). This claim about the reducibility of concepts should express the idea that every empirical concept – including highly theoretical ones such as "electron," "quantum jump," "electric field," "DNA," and, if necessary, also "introverted," "libido," "sublimation," etc. – is either strictly definable in terms of simple observational concepts or is observationally specifiable in an exactly definable weaker sense.

4. TESTABILITY AS VERIFIABILITY OR FALSIFIABILITY

Let us pursue the characterization of empirical sentences a bit further. Claim (1) above, according to which a sentence has empirical content if and only if it is testable by observation reports, obviously has the consequence that sentences not testable in such a way lack empirical content.

A class of sentences that plays a major role in science – namely, the sentences of logic and of mathematics – belongs to the class of nonempirical sentences. It was said of these sentences that they are neither testable by empirical findings nor in need of such testing. For example, the logically true sentence "Tomorrow it will rain in Vienna or it will not rain there" can be proved to be true without any help of weather observations. The meaning of the logical terms *or* and *not* alone is sufficient to secure that any sentence of the form "*p* or not-*p*" is true. Sentences whose truth or falsity could be established entirely independently of any experiential data – exclusively by analysis of the meanings of certain words occurring in them – were then called analytically true or false. In contrast to Mill's radical empiricism, in which logical and arithmetical truths as well were regarded as empirically founded, the empiricism of the Vienna Circle saw those truths as analytical and devoid of empirical content.

But not all sentences that are not testable by observational reports can be considered analytical. According to logical empiricism, this is true especially for certain metaphysical claims, but it also holds true for some untestable systems of ideas that were advanced within empirical sciences (e.g., Driesch's neovitalism). Such systems of sentences were then called cognitively meaningless "pseudohypotheses." In particular, the lack of progress in philosophy was explained by the fact that many classical and contemporary philosophical theses could, in principle, not be critically examined by logical and empirical means. This idea was presented and put to critical use in Carnap's study *Pseudoproblems in Philosophy: The Heteropsychological and the Realism Controversy.*[3]

The little book *Language, Truth, and Logic,*[4] published by A. J. Ayer in 1936 after his studies in Vienna, was an inspiring, though not always sufficiently thought-through, attempt to apply these basic ideas of the Vienna Circle to a broad range of philosophical questions. It played an important role in making these ideas known in English-speaking countries.

But when Ayer's book was published, these ideas had already gone through a process of substantial refinement. This continual refinement and the increasingly liberal construal of the basic ideas of logical empiricism were mainly a result of Carnap's efforts to find a precise definition or "explication" for the notion

---

[3] Rudolf Carnap, *Scheinprobleme in der Philosophie: Das Fremdpsychische und der Realismusstreit* (Berlin-Schlachtensee: Weltkreis, 1928), in Rolf A. George, trans., *The Logical Structure of the World and Pseudoproblems in Philosophy* (Berkeley: University of California Press, 1967), pp. 301–43.

[4] A. J. Ayer, *Language, Truth, and Logic* (London: Gollancz, 1936).

of testability of a phypothesis by observation sentences, which has here been presented only in an intuitive way.

Schlick, for example, at first proposed an idea about testability which was without doubt influenced by a claim Wittgenstein made in the *Tractatus* – namely, that a sentence has to be a truth function of elementary sentences. Schlick thought that every empirically meaningful sentence must, in principle, be verifiable by appropriate observational reports in a full sense (i.e., it must be possible to show that it is true). This so-called criterion of verification for empirical sentences ran into fundamental difficulties. One of them was the fact that laws of nature do not satisfy this condition. But to find laws of nature is one of the main objectives of scientific research, and, therefore, laws of nature certainly are empirical statements. Laws of nature have the form of universal sentences, and they refer to an unlimited class of possible occurrences of single events, which makes it obvious that they cannot be fully verified by a limited number of observations. The criterion of verification was abandoned because of this and other difficulties.

This example shows us how important it is for an adequate explication of the testability of scientific hypotheses to be totally clear about the different logical forms such hypotheses might have. For this reason Carnap and others used in their logical analyses the concepts and methods of the new mathematical logic, which, in terms of subtlety and strength, was highly superior to classical logic, and the importance of which, for the analysis of philosophical problems, was strongly emphasized by logical positivism.

In his important book *The Logic of Scientific Discovery*[5] and in many subsequent publications, Karl Popper proposed and defended a very plausible alternative to the criterion of verification for empirical sentences. The basic idea of this so-called criterion of falsifiability is the following: In general, a scientific hypothesis is tested by first deducing predictions from it in the form of observation sentences that predict the results of relevant

---

[5] Karl Popper, *The Logic of Scientific Discovery* (London: Hutchinson, 1959). This is Popper's translation of his *Logik der Forschung* (Vienna: Julius Springer, 1935).

experiments and then confronting them with the actual findings of observation and experiment. If such a confrontation shows that one of the derived observation sentences is false, then the hypothesis itself is shown to be false. Accordingly, the criterion of falsifiability requires that an empirical hypothesis be, in principle, refutable by observational findings (i.e., it must be possible to describe conceivable results of observation that contradict the hypothesis, and if these observations actually occurred, then the hypothesis would be refuted or "falsified").

This new criterion has the important consequence that universal sentences, which the verification principle disqualifies, are admitted as empirical. For example, the hypothesis "All swans are white" is an empirical hypothesis, because it could be falsified by the observation sentence "This is a not-white swan." Nevertheless, the falsifiability principle is too narrow, and the reasons for that are analogous to those for the verification principle. But this is not the place to give a more detailed explanation.

With reference to the falsifiability requirement, Popper denied that the individual psychology of Alfred Adler and the psychoanalytic system of Sigmund Freud are scientific theories. His reason was that both systems of ideas are untestable and irrefutable: Any imaginable behavior could be interpreted in either and would therefore "confirm" both systems.[6]

Popper rightly points out that adherents of the theories just mentioned are careless in postulating "explanations" of certain cases of human behavior that certainly cannot be derived deductively from the theory. But the following has to be considered with regard to his claim that those theories are not falsified by any imaginable human behavior and are, therefore, confirmed by any imaginable human behavior: If a psychological theory is unfalsifiable in a strict sense (i.e., no statements about observable human behavior can be derived from it that might possibly turn out to be false, thus falsifying the theory), then this theory does not imply any statements about observable human behavior. Accordingly, no imaginable human behavior "confirms" the

---

[6] Compare Karl Popper, *Conjectures and Refutations* (London: Routledge and Kegan Paul, 1963), pp. 33–9.

theory; the theory is not "confirmed" by any imaginable human behavior, let alone by every human behavior. The theory explains not everything, but nothing.

It requires detailed logico-systematic inquiries to find out whether some particular version of a psychoanalytic theory can be falsified. Problems of this sort have been explored by psychologists and philosophers of science for a long time. Such studies, which have refined the relevant questions considerably, show, in any case, the fruitfulness of the empirical requirement that scientific theories have to be testable.

In view of the deficiency of both the verifiability and the falsifiability criterion, Carnap in his important essay "Testability and Meaning"[7] constructed a much more general criterion of confirmability for empirical sentences, which includes the two criteria just mentioned as special cases. But I will not give even a sketchy view of it, because by now we know that all so-far-mentioned attempts to formulate precise criteria for empirical sentences are subject to a fundamental objection that made it necessary to change the formulation of the problem in a radical way.

5. DUHEM: HOLISM

All the analytical attempts we have looked at so far aimed at criteria for empirical testability and for the empirical status of single statements. But it became apparent that the whole statement of the problem rests on a wrong presupposition. Single statements are, in general, not testable by observation at all. This was clearly shown in 1951 by W. V. O. Quine in his penetrating critical essay "Two Dogmas of Empiricism."[8] It is remarkable that already in 1906 Pierre Duhem, in his *La Théorie Physique*,[9] presented the basic

[margin annotation: Duhem 1906]

---

[7] Rudolf Carnap, "Testability and Meaning," *Philosophy of Science* 3 (1936), 419–71, and 4 (1937), 1–40.

[8] W. V. O. Quine, "Two Dogmas of Empiricism, "*Philosophical Review* 60 (1951), 20–43.

[9] Pierre Duhem, *La Théorie Physique: Sun Objet et sa Structure* (Paris: Revière, 1906, 1914); Philip P. Wiener, trans., *The Aim and Structure of Physical Theory* (Princeton: Princeton University Press, 1954).

idea in a clear way, justified it thoroughly, and investigated its consequences.

The idea I am talking about, which is nowadays often called the Duhem-Quine thesis, is the following: A single scientific hypothesis does not normally imply any observation statements by which it could be tested. In order to derive observation statements or experimental results from a singular hypothesis, one has to use many so-called auxiliary hypotheses – among them, for example, the assumptions that no malfunction in the experimental apparatus occurred and that the theoretical assumptions on which the use of the experimental apparatus is based are correct. Accordingly, all these auxiliary hypotheses are additional premises that allow one to derive predictions about experimental findings from the given hypothesis. If a predicted experimental finding is not obtained (i.e., if the derived observation statement has been shown to be false), then this shows only that at least one of the premises used in the derivation has to be false. But by no means is this required to be the hypothesis under investigation; it might also be the case that one of the auxiliary hypotheses is false. The history of science provides us with many examples of this latter scenario.

When an observation sentence that has been derived from a given hypothesis turns out to be false, this does not show that we have to reject the hypothesis as falsified. What it shows is only that somewhere in the whole system of hypotheses that was used in the derivation, some appropriate changes have to be made. But purely logical considerations cannot determine which sentences of that system have to be changed.

But how should it be decided where in the whole theoretical system a change has to be made – and in what way – when we find a "falsifying" experiment? Duhem says there are no strict rules: The decision is made by the scientist's subtle intuition, "le bon sens." Quine treats this question in more detail. He emphasizes certain holistic considerations that play an important role in the scientific decision on how best to change a troubled theoretical system. The new system arrived at by modifications of the starting system should, for example, be as systematically organized as possible; it should be as simple as possible; and it should be in as much accordance as possible with the retained observational findings.

Thomas Kuhn in his historically and sociologically oriented presentation of scientific research takes a similar line in respect to the considerations that influence decisions about which of two conflicting comprehensive theories should be preferred.

These new ideas evidently diverge to a considerable extent from the earlier forms of empiricism developed by the Vienna Circle. The Vienna Circle focused its analyses in the beginning on single empirical hypotheses and tried to characterize their empirical content and their testability with reference to so-called observation sentences, which could be decided by immediate experience. The new view in the spirit of Duhem and Quine is holistic: Empirical content is ascribable only to whole systems of hypotheses, and the decision about whether to accept such a system does not depend only on observational findings but to a large extent also on holistic properties of the system in question.

In consequence, the new view also broke with another central idea of the Vienna Circle. Reichenbach, Putnam, Quine, and others advanced the claim that in order to modify a troubled system, we may have to consider, if more modest measures prove to be insufficient, a revision of the basic principles of logic and mathematics that are built into the theory. Several logicians and philosophers have proposed specific revisions of this very drastic sort for quantum mechanics, for example. But so far these revisions have remained largely unaccepted in physics.

The idea of the revisability of logic and mathematics is obviously incompatible with the view held in the Vienna Circle that the principles of logic and mathematics have a nonempirical and purely analytic status. The new view bears a certain affinity to Mill's empiricist conception of logic and mathematics, although it is refined to fit holism. Naturally it is not claimed that single sentences like "5 + 7 = 12" can be empirically tested and possibly falsified.

283

6. THE FOUNDATION OF KNOWLEDGE

In view of the different possibilities of resolving conflicts between a theoretical system and empirical observations, another question becomes important, a question that was already discussed by Neurath, Carnap, and Schlick in their years in Vienna. The question I have in mind is whether the observation sentences that are based on immediate experience and that are used for testing a system of hypotheses can be said to be true by reference to the relevant immediate experiences – that is, whether they are the unchangeable and unshakable foundation of knowledge.

Schlick, in his essay "On the Foundation of Knowledge,"[10] answered this question in the affirmative, at least in a certain weakened way, and for a time Carnap leant toward the view that observation sentences that are based on immediate experience are neither in need of nor capable of further testing.

Against that, Popper and Neurath – and after a while Carnap as well – held the view that observation sentences, like all other empirical sentences, are open to further testing and, in the end, are also open to the possibility of being rejected. Therefore, they are in principle revisable, and for this reason they cannot form a solid, unchangeable foundation of knowledge.

Popper summarized his views on the basis of empirical knowledge in a striking metaphor: the edifice of scientific knowledge is not built on a rock bottom, but it rests on piles seated in a swamp. As long as the piles support the building sufficiently, one does not pay attention to how reliable this foundation is. But when the whole structure loses stability, the posts are driven deeper into the swamp, until they can carry the building again.

Neurath used a metaphor as well, but one that, in my opinion, expresses a more radical view. In our search for reliable knowledge we are, he says, like sailors who always sail on the open sea and who have to secure and to improve the seaworthiness of

---

[10] Moritz Schlick, "On the Foundation of Knowledge," Peter Heath, trans. in Henk L. Mulder and Barbara F. B. Van de Velde-Schlick, eds., *Philosophical Papers*, vol. 2 (Dordrecht: Reidel, 1979), pp. 370–87. This essay was first published as "Über das Fundament der Erkenntnis," *Erkenntis* 4 (1934), 77–99.

their ship by constantly repairing and rebuilding their ship with materials that are either on board or found floating in the sea. There is no possibility of bringing the ship into a drydock and totally rebuilding it on solid ground. Here you find no foundation of knowledge at all. Not only is there no rock, there is not even a swamp. [QUOTE!]

Skeptical puzzlement about the complete absence of an empirical basis in Neurath's conception of knowledge might be softened by a nowadays widely accepted extension of Duhem's holism: When a conflict appears between a theoretical system and experimental findings, the conflict can sometimes be removed in a plausible way without any change of the system. One can just reject the observation reports. This can definitely be a rational procedure. The history of science offers many examples in which isolated reports of observations that contradict an otherwise well supported theory have been set aside for this very reason. And in everyday life we react in a similar way: If someone told us that he had seen a man flying around St. Stephan's Tower for two hours just by moving his arms, we would reject this report simply because it is incompatible with thoroughly and variously supported information we have about such possibilities. [cite!]

As an example from science we could mention the long dispute between the experimental physicist Felix Ehrenhaft of Vienna and the American physicist Arthur E. Millikan about the correctness of the data on which Millikan based his hypothesis that there is an elementary electrical quantum (the electron) and that its charge is of a certain amount. Ehrenhaft repeated Millikan's experiment many times, but the data he got were incompatible with Millikan's hypothesis, so Ehrenhaft rejected this hypothesis. But Millikan's hypothesis proved to be theoretically very fruitful, and it was also supported by manifold other experiments. Millikan won the Nobel Prize, and eventually no one paid attention anymore to Ehrenhaft's experimental results. Interestingly enough, a new study based on Millikan's laboratory notes in Gerald Holton's book *The Scientific Imagination*[11] shows

[11] Gerold Holton, *The Scientific Imagination: Case Studies* (Cambridge, UK: Cambridge University Press, 1978). See also Chapter 12, pp. 244–45, here.

that Millikan himself encountered some diverging measurement results, but he attributed these results to undetermined interference factors and did not include them in his published findings.

Obviously, extending Duhem's holism radically changes the narrower empiricist thesis that observation sentences can be advanced on the basis of immediate experience independently of any theoretical presuppositions, and can then serve as last authority in testing theories. The opposite is the case: All observation reports have theoretical presuppositions and implications. In case of a conflict between theory and observation reports, sometimes successful theories decide what is a reliable observation report.

It is certainly remarkable that W. V. O. Quine, one of the most influential and most constructive critics of logical empiricism, chose Neurath's radically holistic metaphor, almost thirty years after its publication, as a motto for his book *Word and Object*,[12] which was dedicated to another thinker of the Vienna Circle – namely to Rudolf Carnap.

Schlick, at this time, was much less impressed by Neurath's views than Quine later was. Schlick accused Neurath of advancing a coherence theory of knowledge, according to which the truth of a system of empirical statements is simply determined by the logical coherence of the system, independent of any experiential data.

It is certainly correct to say that in science, a theory is not yet regarded as well justified if its sentences only form a logically coherent system. Even fairy tales can be coherent systems of statements, but they do not for this reason count as scientifically acceptable descriptions of certain parts of the world. What distinguishes a good theory from a fairy tale is the theory's objectivity or intersubjectivity: Sufficiently qualified scientists agree within certain limits on the claims such theories make.

This intersubjective agreement, without which science would be impossible as a social enterprise, certainly depends on a far-reaching agreement with respect to the findings of observation and experiment – here lies the empirical character of science. But,

---

[12] W. V. O. Quine, *Word and Object* (New York: Wiley, 1960).

as we have seen, this agreement cannot be explained in a satisfactory way by postulating a basis of theoretically neutral observation sentences common to all humans. We have to look for a better explanation within the scope of a genetic-causal theory of human observation, perception, and hypothesis formation. For quite some time, philosophy has worked on the beginnings of such a theory. Quine, for example, has dealt with this problem in a series of papers. One of them has the characteristically programmatic title "Epistemology Naturalized." And nowadays a whole group of younger researchers tries to build an empiricist-pragmatist epistemology.

This pragmatic turn is as alien to the philosophy of science of the Vienna Circle as is the new holistic view of knowledge, the rejection of the idea that the theory-neutral findings of observation can ground all empirical knowledge as well as the refusal to accept a dichotomy of analytic and synthetic sentences. But this way of introducing refinements or even radical revisions of formerly guiding ideas is analogous to the metamorphosis of scientific theories. In a scientifically-oriented philosophy as well as in the empirical sciences, the path of progress is paved with the fragments of revised or totally rejected ideas; and this is exactly in accordance with the ideal of "scientific philosophy" held by the Vienna Circle.

Chapter 15
Hans Reichenbach Remembered*

Hans Reichenbach's far-ranging and influential contributions to epistemology and the philosophy of science will be acknowledged and appraised in many contexts on the occasion of the one-hundredth anniversary of his birth. The following pages, however, are simply meant to record some personal recollections of Reichenbach as he affected one of his many students.

Reichenbach came to the University of Berlin in 1926, over strong philosophical opposition, but with stronger scientific support, especially by Einstein. I had by then been a student there for a year, after previous studies in Göttingen and Heidelberg. I promptly enrolled in a course with Reichenbach and was startled by his ideas on causality and determinism, which contrasted sharply with what I had been taught earlier, especially in courses on Kant's *Critique of Pure Reason.* Early on, I told him that I considered the principle of causality as true a priori and thus found it impossible even to imagine that it might be false. With a characteristic smile – and an aside on how limited one's imagination could be – Reichenbach said that evidently I had swallowed Kant's conception hook, line, and sinker. But not to worry: He was confident I would change my mind in light of the arguments he would be presenting. I did indeed change my mind on this subject and on many others; Reichenbach's ways of exploring

* From Carl G. Hempel, "Hans Reichenbach Remembered," *Erkennthis* 35 (1991), 5–10. Reprinted with the kind permission of Kluwer Academic Publishers.

philosophical issues came to exert a strong influence on the development of my thinking and led me into the arms of empiricism.

Reichenbach tended to be quite sure of his ideas, and occasionally he was too readily dismissive of dissenting views, but this did not affect his refreshingly receptive and informal way with students. At the time, many of the outstanding professors at German universities were virtually unapproachable to their students, specially beginners. I recall once running after Max Planck to ask whether I might put a brief question to him about a point he had made in his lecture. He did not even turn around. "Ask my assistant," he said, vanishing into his office. That was not rudeness or arrogance; it simply wasn't the job of a professor to bother with such questions.

It was quite different in Reichenbach's classes – and, happily, in those of some of my other academic teachers, among them Paul Bernays in Göttingen, Wolfgang Köhler and Kurt Lewin in Berlin, and many members of the Vienna Circle.

Reichenbach had a felicitous way of giving his students a sense of being engaged in a joint effort to solve problems of importance and even of being potential contributors to progress in the field. I am quite sure that he did not always have an answer to the questions he encouraged his students to explore. One of the seminars I took with him was devoted to an attempt at formalizing Hilbert's axiomatization of Euclidean geometry in the notation of *Principia Mathematica*. We encountered great difficulties with the axiom of completeness – Reichenbach himself did not know whether it could be formalized with the contemplated means. We tried hard and eventually concluded, I think, that it was indeed impossible.

But Reichenbach's interaction with his students was not restricted to the classroom. He took a strong interest in some of the personal problems, the plans, hopes, and doubts of his students, and occasionally he joined some of them on an excursion to the lakes and woods in the environment of Berlin. My friend and fellow student Olaf Helmer, who took several of Reichenbach's courses, has spoken to me with much appreciation of this aspect of Reichenbach's sympathy and concern for his students.

Reichenbach's seminars sometimes adjourned to a nearby café, where broader issues of social and political concern were debated as well. The café also provided a welcome setting for ideological and often political conversations among Reichenbach and some of his colleagues at the *Gesellschaft für empirische Philosophie*, including Walter Dubislav (of the Technische Hochschule) and Kurt Grelling, the discoverer of the paradox concerning the term *impredicable*, a gifted mathematician and philosopher who taught at a high school in Berlin. This kind and humane man, who was of Jewish ancestry, was to die later in the gas chambers of Auschwitz.

The *Gesellschaft für empirische Philosophie* – in contrast to the Vienna Circle, which was a small closed discussion group of scholars – imposed no membership restrictions. Reichenbach, Dubislav, and Grelling were the leading figures. The group organized, each academic year, a series of lectures, open to the general public, which were held in the large, amphitheatrical auditorium of the *Charité*, the University Medical School. The lectures covered a wide range of topics, reflecting a distinct openness to diverse issues and approaches. The speakers included traditional philosophers as well as members of the Vienna and the Polish groups of methodologists and logicians, among them Carnap, Tarski, and Neurath. Some of the lectures dealt with psychoanalysis and other topics in psychiatry; the Berlin psychiatrist Alexander Herzberg was among the more influential members of the *Gesellschaft*. But I recall also a lecture-performance of an academically quite unorthodox kind. It was given by a man offering to duplicate and explain the various feats of a psychic then much in the public eye, a person named Hanussen. The occasion attracted an audience much larger than even the *Charité* could accommodate. I recall Dubislav, who could be quite brusque, scurrying about and urging the persons sitting on the steps between the aisles to leave voluntarily in conformance with fire-safety regulations – or would they rather have him call in the police? The speaker performed and explained a number of feats, such as finding hidden objects and reading the thoughts of persons in the audience.

In a letter to a fellow-student, date November 6, 1929, I wrote about a lecture that Carnap gave at the *Gesellschaft für empirische Philosophie* on questions concerning Wittgenstein's logical atomism. The discussion lasted for four hours, the final two of them at a café, where the intensely involved participants – among them Reichenbach, Scholz, Bernays, Kurt Lewin, Grelling, Dubislav – became so agitated and noisy that they practically caused a public nuisance and made young couples at neighboring tables break off their tender exchanges.

The *Gesellschaft für wissenschaftliche Philosophie* in cooperation with its sister-organization, the *Verein Ernst Mach* in Vienna, sponsored the publication of the periodical *Erkenntnis*, which succeeded the *Annalen der Philosophie*, with Carnap and Reichenbach as editors. The first issue, which bears no publication date, must have appeared in 1930. Reichenbach wrote the introduction and played a leading role in guiding the further development of the journal.

The first issue contained articles by Schlick, Carnap, Dubistlav, and Reichenbach. They clearly reflected what Schlick, in the title of his contribution, called "Die Wende der Philosophy." The next issue of volume 1 was devoted to a "Report on the first convention on the epistemology of the exact sciences"; subsequent numbers addressed philosophical questions concerning biology, and the first issue of volume 2 included an article by the astronomer Erwin F. Freundlich on the question of the finitude of the universe, treated as an astronomical problem. All this bespoke a keen concern to avoid very general philosophical discussions, always seen as threatening to slide into "meaninglessness," in favor of problems concerning specific aspects of scientific inquiry.

In 1928, I became acquainted with Carnap's *Der logische Aufau der Welt* and his less technical writings criticizing metaphysics. I decided to study in Vienna for a term – a plan that was reinforced when I met Carnap in person at the first *Tagung für die Erkenntnislehre der exakten Wissenschaften* in Prague, 1929, which was an important milestone in the development of logical empiricism. Reichenbach supported my idea, and his letter of introduction to

Schlick produced an immediate invitation to join the discussions of the Vienna Circle, which, together with various courses I attended, had a lasting effect on my philosophical development. It was not long after my return from Vienna, I think, that Reichenbach introduced me to Paul Oppenheim, an acquaintance of his who was a chemist by training but who had also published a book and some articles on logical facets of concept formation in science. He wanted a critical appraisal of his ideas and advice on ways to improve and expand his conceptions. That was the beginning of a collaboration that was to continue for many decades, a collaboration dealing with a variety of topics beyond those that started it; it was also the beginning of a deep friendship. With the advent of National Socialism, Oppenheim moved to Brussels; at his invitation, I joined him there in 1934. This made it possible for me to leave Germany and, with Carnap's help, to go to the United States in 1937. I hate to think of what might have become of me if Reichenbach had not put me in touch with Oppenheim.

After my return from Vienna, I began work on my doctoral dissertation, which dealt with the statistical concept of probability. Reichenbach agreed to be my thesis adviser (*Doktorvater*), and I began work with him. But in 1933 he was relieved of his position in Berlin because of his Jewish ancestry. He promptly accepted a professorship at the University of Istanbul, where he maintained his professional links to *Erkenntnis* and other enterprises by correspondence. He joined the faculty of the University of California in Los Angeles in 1938.

Reichenbach's dismissal made it necessary for me to seek new sponsors and examiners for my dissertation project; I am grateful to Wolfgang Köhler, with whom I had done course work, and to Nicolai Hartmann, who did not know me at all, for having agreed to take on that role. I completed my dissertation on problems of the frequentist conception of probability on my own. But Reichenbach, who had introduced me to the subject, showed great interest in what I had written, and, although I had expressed reservations about details of his own view, he saw to it that a condensed version of my study was published in *Erkenntnis* in 1935. He added a response of his own, welcoming some of my ideas but rejecting my critical doubts about his own conception of the subject.

The files of my correspondence with Reichenbach run from 1929 to 1951. They begin with an account I gave him in December 1929 of my first impressions of Vienna and especially of the discussions in the Vienna Circle and of the courses I was taking with various philosophers at the university. Reichenbach had expressed a keen interest in these matters, and I wrote him at his request. Our subsequent correspondence ranged over many topics, among them editorial matters concerning *Erkenntnis*; requests for me to write reviews and to do extensive proofreading – the latter also for some of his books; discussions of each other's manuscripts; information about mutual friends, including Grelling, for whom Reichenbach wrote letters supporting his efforts to be admitted to the United States. In a letter written in December 1933, when Reichenbach was already in Istanbul, I reported to him about a meeting of the *Gesellschaft für empirische Philosophie* at which Dubislav had lectured on the laws of nature. The meeting, I said, was less well attended than on earlier occasions, but the level of the discussion was still good. I also gave him accounts of disturbing events set in motion by National Socialism; the eminent mathematician Ludwig Bieberbach strutting about in a Nazi uniform, greeting his classes with the Hitler salute, and talking of the racial facets of mathematics.

Initially, the language of our correspondence was German, of course; in the mid-1940s we changed to English. Reichenbach's first English letter informed me that he had recommended me for a vacancy at Vanderbilt University. Nothing came of the recommendation, but I had by then obtained a satisfying position at Queens College in New York.

In 1951, when I spent some time in Los Angeles, Reichenbach immediately offered to put me up at his house; when this proved impractical for me, he and his wife extended warm hospitality to me during a short visit.

Reichenbach's letters to me were predominantly concerned with philosophical and editorial issues. This was so even when he had gone to Istanbul, which meant a radical change in his life. He did, however, make a few incidental remarks about his daily life and his teaching experiences. In November 1933, for example, he writes that he has to keep his lectures rather elementary, but hopes to be able gradually to raise the level of his courses. The interest of his students, he says, is touching, so that his work gives him much pleasure. He remarks on the surprises awaiting him and his wife on visits to mosques and bazaars. Once, they had just been to a bazaar and asked for a camel hair blanket for their bed, whereupon the merchant brought them a heavy cover of the kind put on a camel's back for riding on it.

In subsequent letters, he mentions having a whole row of assistants with whom he can alternately speak German, French, and English – but whose philosophical training is rather weak. He also tells of lecturing on the history of philosophy, dealing with the thinkers in question more from a psychological and sociological perspective, which he finds quite an interesting mode of approach. Additionally, he refers to a colloquium held by a small circle of scholars speaking German. But, he adds, that group is only a weak substitute for the circle in Berlin.

In all, I remember Hans Reichenbach as a powerfully creative philosophical and scientific thinker and writer, as a dedicated and enlightened educator, and as a man deeply concerned with the human condition and dedicated to its enhancement. He was a *mensch*.

Chapter 16

Empiricism in the Vienna Circle and in the Berlin Society for Scientific Philosophy
Recollections and Reflections*

1. INTRODUCTORY REMARKS ON THE TWO EMPIRICIST GROUPS

The central ideas of logical, or scientific, empiricism as it developed during the twenties and early thirties in Vienna and in Berlin grew out of collaborative efforts of scientifically interested philosophers and philosophically interested scientists. Those thinkers noted that while the claims made by the physical sciences were amenable to objective test by experiment and observation, the pronouncements put forward by metaphysics were incapable of any such objective critical appraisal. And while hypotheses advanced in the physical sciences would eventually be accepted or rejected and thus lead to the growth of a body of objective scientific knowledge, the problems and pronouncements of metaphysics, inaccessible to objective appraisal, kept reappearing over and over again.

Moritz Schlick, the founding father of the Vienna Circle, addressed this situation in an article entitled "Die Wende der Philosophie," which appeared in 1930 in the first issue of the empiricist periodical *Erkenntnis*. Rejecting "the anarchy of philo-

* From Carl G. Hempel, "Empiricism in the Vienna Circle and in the Berlin Society for Scientific Philosophy: Recollection and Reflection," in Friedrich Stadler, ed., *Scientific Philosophy: Origins and Developments* (Dortrecht: Kluwer Academic Publishers, 1993), pp. 1–9. © Kluwer Academic Publishers. Reprinted with the kind permission of Kluwer Academic Publishers.

sophical opinions," he solemnly declared, "We stand in the middle of a definitive turn of philosophy and are justified in regarding the fruitless strife of philosophical opinions as finished."[1] The means for avoiding all such fruitless strife he saw in the clear and precise thinking exemplified by the exact sciences and informed by the principles of exact formal, or symbolic, logic.

Broadly speaking, the Vienna Circle held that the purported problems of metaphysics constitute no genuine problems at all and that in an inquiry making use of an appropriately precise conceptual and linguistic apparatus, metaphysical questions could not even be formulated. They were pseudoproblems, devoid of any clear meaning.

Some of the logical empiricists – especially Neurath and Carnap – also noted the potential for political and religious misuse of metaphysical discourse.

In the preface to his otherwise highly technical work, *Der logische Aufbau der Welt*, Carnap offers some brief but fascinating remarks that place the ongoing revolution in philosophy into a wider cultural context:

> We feel an inner affinity between the attitude that lies at the bottom of our philosophical work and the spiritual attitude which expresses itself at present in entirely different areas of life: we sense that attitude in currents of (contemporary) art, especially in architecture, and in the movements that seek to give a meaningful shape to human life.[2]

In this spirit, Carnap took a lively interest in the ideas developed at the Bauhaus. He gave lectures there (as did Neurath and Feigl)

---

[1] Moritz Schlick, "The Turning-Point in Philosophy," Peter Heath, trans., in Henk L. Mulder and Barbara F. B. van de Velde-Schlick, eds., *Moritz Schlick: Philosophical Papers*, vol. 2 (Dortrecht: Reidel, 1979), pp. 154–60. This essay first appeared as "Die Wende der Philosophie," *Erkenntnis* 1 (1930), 4–11.

[2] From Hempel's Preface to Rudolf Carnap, *The Logical Structure of the World*, Rolf A. George, trans. (Berkeley: University of California Press, 1967). This book was first published as *Der logische Aufbau der Welt* (Berlin-Schlachtensee: Weltkreis, 1928).

and later at the Chicago School of Design, where he exchanged ideas with L. Moholy-Nagy.

A sense of involvement in a novel kind of philosophical enterprise inspired also the logical empiricist group in Berlin, though there was a less pronounced emphasis on a radical break with the old tradition. Reichenbach, however, did clearly refer to the new turn in his book, *The Rise of Scientific Philosophy*.[3] The first part of that book, he says, is concerned with the shortcomings of the earlier "philosophy of speculation," while the second part presents "modern scientific philosophy," which employs the methods of contemporary logical analysis and aims to "present the evidence that philosophy has risen from error to truth."

3.* THE VIENNA CIRCLE

In 1928, I became acquainted with Carnap's *Der Logische Aufbau der Welt* and his less technical pamphlet "Die Überwindung der Metaphysik durch logische Analyse der Sprache." I promptly decided to study in Vienna for a term – a plan that was reinforced by my meeting Carnap in person at the first *Tagung für die Erkenntnislehre der exakten Wissenschaften* in Prague in 1929, which was a significant milestone in the development of logical empiricism.

Reichenbach supported my idea, and his letter of introduction to Schlick promptly produced an invitation to attend the discussions of the Vienna Circle.

Among the participants in those sessions, I recall especially Schlick, Carnap, Otto Neurath, Hans Hahn, Herbert Feigl, and Friedrich Waismann. As I remember it, Karl Menger and Felix Kaufmann were frequently present and so was the physicist Philipp Frank, Einstein's successor at the University of Prague. Quite a few of the leading members of the Vienna Circle were not philosophers by primary training: Schlick had earned his doc-

---

[3] Hans Reichenbach, *The Rise of Scientific Philosophy* (Berkeley: University of California Press, 1951).

* Section 2 of this article, "The Empiricist Group in Berlin," is a reminiscence that repeats much of the first third of chapter 15, pp. 288–93. (Editor)

torate in physics in Berlin, Hahn and Menger were mathematicians, Neurath a sociologist.

Karl Popper was in Vienna at the time and maintained a lively exchange of ideas with members of the Circle.

But he kept at a definite philosophical distance from the Vienna Circle – a distance which I think was excessive; for after all, there was no party doctrine to which the members of the group were committed. What those thinkers did share was a sense of dissatisfaction at the thought that, in contrast to the natural sciences, philosophy had had so little success in its endeavors to solve certain problems that were widely viewed as deep and important – especially problems of metaphysics. Most members of the Vienna Circle shared the conviction that by means of precise analytic methods, informed by modern logic, such recalcitrant problems could either be solved or shown to be pointless "pseudoproblems." Finally, the thinking of the group was inspired by a basically empiricist conception of knowledge.

At this point, let me call attention to a public "Aufruf" reflecting a similarly empiricist inspiration, which was published in 1912 by a group of eminent mathematicians, scientists, and scholars, among them Albert Einstein, David Hilbert, Ernst Mach, Sigmund Freud, Felix Klein, and Jacques Loeb.[4] That appeal was addressed to "all philosophically interested researchers," including scientists as well as all philosophers, "who expect to arrive at tenable claims only by a penetrating study of the facts of experience itself"; it invited them to join a "Society for Positivist Philosophy," which was to establish live connections between all the sciences and to advance to a consistent comprehensive world view.

To return now to the Vienna Circle: While the members of the group shared an empiricist outlook and the rejection of meta-

---

[4] This document is in the *Zentrales Archiv der Akademie der Wissenschaften*, in the former East Berlin. I am much indebted to my friend, Professor Adolf Grünbaum, at the University of Pittsburgh, who brought this document to my attention and provided me with a copy of it and of a more detailed account of it, which appeared in *The Journal of Philosophy, Psychology, and Scientific Methods* IX, 15 (July 18, 1912), 419–20.

physics as cognitively meaningless, there were considerable differences between them concerning further specific issues, as well as marked differences in personality and scholarly style.

Carnap, introverted, cerebrotonic, and extremely systematic, insisted on formulating philosophical issues as clearly as possible and on exploring them by means of exact logical methods, including the use of contemporary symbolic logic. He formulated his ideas, including his construal of physicalism, in precise terms and modified them by reference to his increasingly refined explorations of scientific usage. Around 1930 he proposed a version of physicalism according to which psychological characteristics such as anger, excitement, fatigue, and hunger are *dispositions* of organisms to behave in characteristic ways that can be described in purely physical terms. This view anticipated, though in a limited and sketchy form, the conception of mental states and events that Gilbert Ryle was to set forth in his important and influential work, *The Concept of Mind* (1945).

Carnap's characteristically thorough, systematic, step-by-step way of tackling philosophical issues was no doubt considered pedantic and ponderous by some others. There was, for example, a profound contrast between Carnap's philosophical style and that of Wittgenstein, who relied much more on a certain intuitive flair in philosophizing. Indeed, in 1928 this conflict in style led Wittgenstein to break off all direct philosophical exchanges with Carnap.

Waismann's ways in his seminar and in conversation were exemplarily thoughtful and thorough. He strongly urged and stimulated his students to think for themselves.

I recall Schlick as a man of aristocratic personality, a bit aloof, yet unfailingly kind and courteous and willing to give consideration to ideas uncongenial to himself, such as Neurath's.

Otto Neurath was a plumpish, jovial man, extroverted, endlessly energetic and enterprising, always bubbling over with ideas and projects. He often signed letters to friends with a self-portrait in the form of a laughing elephant who carried Neurath's initials – an N inside a circular O – on his hindquarter and held in his trunk a bouquet of flowers for the addressee. As for his discursive style, Neurath tended to express his views not so much

in carefully articulated and reasoned arguments as in programmatic pronouncements and calls for action. He maintained a
version of physicalism according to which all things and events
are of a purely physical kind and can be characterized by means
of a physicalistic vocabulary; utterances violating this requirement were regarded as confusion and speculation. Since many of
the hypotheses and theories in the psychological, sociological,
and historical disciplines speak of intentions, beliefs, fears,
hopes, customs, etc. – and thus use terms that are not obviously
physical – Neurath considered it an important task to examine to
what extent and in what manner those locutions could be recast
in a physical language.

Accordingly, he invited his listeners and readers to join forces
in a large effort to separate, among the pronouncement of the
Geisteswissenschaften, the empirical wheat from metaphysical
chaff. To avoid the pitfalls of empty verbiage, he urged that children be trained from the beginning to speak only physicalese, as
it were.

4. METHODOLOGY OF SCIENCE:
NORMATIVE OR DESCRIPTIVE?

Notwithstanding the important agreements that have been
noted, Neurath's ideas differed fundamentally from those of
mainstream logical empiricism as advocated by Carnap. Briefly
and crudely speaking, Carnap construed his methodological
theory as a system of rules and conventions; Neurath regarded
the methodology of science as an empirical, descriptive study of
the research behavior of scientists.

Carnap held that in calling for a logical linkage between
scientific hypotheses and experiential evidence, analytic empiricists meant to offer a *requirement*, a *prescription*, a *norm* for meaningful empirical discourse, not the descriptive assertion that the
claims advanced by scientists are actually always properly linked
to experiential findings. Carnap and Popper (but not Neurath, as
will be seen below) were emphatic in rejecting such a "naturalistic" view. Accordingly, they held it to be strictly irrelevant for

the logical analysis of science to study the biological, psychological, and sociological factors that can affect scientific inquiry as a concrete human activity. There was, to be sure, a polite bow in the direction of a pragmatic study of the psychological, historical, political, and social facets of actual scientific research behavior: such study might shed light on the ways in which that behavior deviates from analytic-empiricist standards.

One member of the Vienna Circle, Otto Neurath, strongly opposed the construal of the methodology of science as aiming at rational reconstructions in precise logical terms. He insisted that an adequate characterization of scientific inquiry had to provide a physicalistic account of the research behavior of scientists, a *Gelehrtenbehavioristik*.

Thus, Neurath rejected Carnap's idea of a rational reconstruction – or explication – of science via terms and sentences, all of which had precisely specified meanings. How, for example, could the meanings of all terms used in a scientific theory be specified precisely? The means for such specification would always have to consist of the terminological apparatus available at the time – for example, right now. But this apparatus, Neurath emphasized, is always a *universal slang*: It consists in part, indeed, of very precise terms and statements such as those used in theoretical physics, but it also contains a large amount of quite fuzzy locutions which we have to use in contexts where a terminological apparatus is *in statu nascendi*.

He referred to such fuzzy terms of colloquial language as *Ballungen* – as clumpings. Their use is inevitable in the early stages of a system of assumptions and claims in an area of inquiry; there is never a point where they can be totally dispensed with. A characterization of all aspects of the world by means of *Laplace's universal formula* couched in precise mathematical notation is a chimera. *Reliance on a universal slang with its fuzzy Ballungen is inevitable in the formulation of our ideas at any stage of scientific inquiry.*

Neurath also proposed a characterization of "tautologies" – logical truths – in pragmatist terms. Consider two orders that Otto might give to Karl:

1. Go outside if the flag is waving;
2. Go outside if the flag is waving and $2 \times 2 = 4$.

The addition of the tautology "$2 \times 2 = 4$" does not alter the effect of the command.

Clearly, this way of giving tautologies a pragmatist interpretation faces considerable difficulties; I mention it to illustrate the rough-and-ready way in which Neurath sought to incorporate statements of logic and arithmetic into his behaviorist-pragmatist conception of scientific claims.

Neurath's presentation of the methodology of science as part of a *Gelehrtenbehavioristik* is limited to examples of the simple kind I have mentioned. He does not address, as far as I am aware, the broader and more technical questions that usually occupy the methodology of science, such as these: Under what conditions do scientists accept a statement, or a theory? When there are two theories for a certain class of occurrences, which, if either, will be accepted in a given situation? Questions of this kind have been dealt with by other thinkers in a pragmatist, but not narrowly behaviorist, tradition, among them C. S. Peirce, William James, John Dewey, and Ernest Nagel. One of Dewey's books is characteristically entitled *How We Think*, not *Rational Thinking*.

Otto Neurath was a pioneer of methodological pragmatism already in the days of the Vienna Circle. It has become clear by now that a pragmatic approach is indispensable for an adequate understanding of scientific inquiry.

As an earlier adherent of the antinaturalist approach to the methodology of science, I have been particularly impressed – and influenced – by Thomas Kuhn's extensive studies of scientific research from a historical, sociological, and psychological perspective; his work makes a powerful case for the indispensability of a pragmatist approach to the subject.

The empiricism of the Vienna Circle thus was gradually liberalized as a result of continuing critical reappraisal by members of the group and by others, the most important among them being W. V. Quine, who, in "Two Dogmas of Empiricism" and in his book *Word and Object*, undermined some of the basic truths of logical empiricism. In the course of this critical development, the early foundationalist conception of a firm basis of experiential data came to be replaced by a coherentist – or holistic – conception of scientific claims as forming a system of mutually interdependent assertions, some of them accepted on the basis of immediate observation, others inferentially linked to them and to one another, all offering mutual support to each other, but every one of them always open to reconsideration and change.

Popper characterized this conception by his often-cited metaphor that the edifice of science does not rest on a firm, rock-bottom foundation but is, rather, supported by pillars that reach down into a swampy ground. As long as the construction of the superstructure encounters no difficulties, the question of its foundations does not receive much attention. But when cracks appear in the edifice, efforts are made to drive the pillars more deeply into the swampy bottom, so that they can support the building again.

Neurath invoked a different metaphor, which was meant also to reflect his misgivings about Carnap's idea of an ideal language of science, one with a precise logical articulation: In our search for knowledge, he said, we are like seafarers who are obliged forever to rebuild their ship on the open ocean, using materials that float by, without ever being able to take the craft into a drydock to rebuild it from the bottom up on a firm basis.

It seems significant to me that nearly three decades later, W. V. Quine, one of the most influential and constructive critics of logical empiricism, chose Neurath's metaphor as a motto for his work *Word and Object*, which he dedicated to Carnap.

Thus, under the influence of internal and external critical and constructive reflection, the original ideas of the Vienna Circle evolved into a new and very different empiricist construal of scientific knowledge and of the methodology of scientific inquiry.

While surely not initially intended or anticipated, this evolution was in accord with a basic view widely held in the Vienna Circle as to how a "scientific world view" was to be developed: not as a monolithic a priori system, but, like empirical science itself, under constant critical appraisal and revision.

303

5. EPILOGUE

The two vigorous empiricist groups we considered at the beginning were destroyed by National Socialist and kindred political powers.

Reichenbach, who was of Jewish ancestry, was relieved of his post in Berlin in 1933. He joined the faculty of the University of Istanbul, where he maintained close contacts with his empiricist colleagues elsewhere, dealing very actively with philosophical and organizational matters. In 1938, he came to the United States to join the faculty of the University of California in Los Angeles. He died in 1953. Dubislav left for Prague, where he took his own life in 1937. Schlick was assassinated in 1936 by a student who claimed that Schlick's positivism had undermined his *Weltanschauung*. Grelling, a very gentle, highly principled man, perished with his wife in the gas chambers of Auschwitz. Neurath escaped first to Holland and then to England, where he vigorously continued his work until his death in 1945.

Other members of the empiricist groups in Berlin and Vienna, among them Carnap, Feigl, Frank, Menger, and myself, had the good fortune of finding new and challenging opportunities to continue their work in the United States. There, they came to form close ties to American empiricist thinkers such as Charles Morris, who had a more pragmatist orientation inspired by John Dewey.

After World War II, the ideas of logical empiricism, especially in the form developed by Carnap, were extensively presented and critically discussed in Germany by Wolfgang Stegmüller at the University of Munich and by a number of his students who came to occupy chairs at several German universities. And it was Stegmüller and his former student Wilhelm K. Essler at the University of Frankfurt who proposed and successfully implemented the idea of resuming the publication of *Erkenntnis*, which had been shut down in 1939. The "new" *Erkenntnis* started in 1975 with volume 9 and is thriving to this day.

C. G. Hempel's Publications

[1] *Beitraege zur logischen Analyse des Wahrscheinlichkeitsbegriffs,* Jena: Neuenhahn (Ph.D. dissertation, Berlin: 1934), 72 pp.

[2] "On the Logical Positivists' Theory of Truth," *Analysis* 2 (1935), 49–59.

[3] "Analyse logique de la psychologie," *Revue de Synthese* 10 (1935), 27–42. (English translation, by Wilfred Sellars, under the title "The Logical Analysis of Psychology," in H. Feigl and W. Sellars, eds., *Readings in Philosophical Analysis.* (New York: Appleton-Century-Crofts, Inc., 1949), pp. 373–84.

[4] "Zur Frage der wissenschaftlichen Weltperspektive," *Erkenntnis* 5 (1935/36), 162–4.

[5] "Über den Gehalt von Wahrscheinlichtkeitsaussagen," *Erkenntnis* 5 (1935/36), 228–60.

[6] "Some Remarks on 'Facts' and Propositions," *Analysis* 2 (1935), 93–6.

[7] "Some Remarks on Empiricism," *Analysis* 3 (1936), 33–40.

[8] With P. Oppenheim, *Der Typusbegriff im Lichte der Neuen Logik: Wissenschaftstheoretische Untersuchungen zur Konstitutionsforschung und Psychologie* (Leyden: A. W. Sijthoff, 1936).

[9] With P. Oppenheim, "L'importance logique de la notion de type," *Actes du Congrés International de Philosophie Scientifique,* vol. II, (Paris: Hermann et Cie, 1936), pp. 41–9.

[10] "Eine rein topologische Form nichtaristotelischer Logik," *Erkenntnis* 6 (1937), 436–42.

[11] "A Purely Topological Form of Non-Aristotelian Logic," *The Journal of Symbolic Logic* 24 (1937), 97–112.

[12] "Le problème de la vérité," *Theoria* 3 (1937), 206–46. Translated by the author as "The Problem of Truth."

[13] "Ein System verallgemeinerter Negationen," *Travaux du 19e Congrès International de Philosophie*, vol. VI, (Paris: Hermann et Cie, 1937), pp. 26–32.

[14] "On the Logical Form of Probability-Statements," *Erkenntis* 7 (1938), 154–60.

[15] "Transfinite Concepts and Empiricism," *Synthese* 3 (1938), 9–12, in the section entitled "Unity of Science Forum."

[16] "Supplementary Remarks on the Form of Probability Statements," *Erkenntnis* 7 (1939), 360–3.

[17] "Vagueness and Logic," *Philosophy of Science* 6 (1939), 163–80.

[18] Articles "Whole," "Carnap," "Reichenbach," in D. Runes, ed., *Dictionary of Philosophy* (New York: Philosophical Library, 1942).

[19] "The Function of General Laws in History," *The Journal of Philosophy* 39 (1942), 35–48.

[20] "A Purely Syntactical Definition of Confirmation," *The Journal of Symbolic Logic* 8 (1943), 122–43.

[21] "Studies in the Logic of Confirmation," *Mind* 54 (1945), 1–26 and 97–121.

[22] "Geometry and Empirical Science," *The American Mathematical Monthly* 52 (1945), 7–17.

[23] Discussion of G. Devereux, "The Logical Foundations of Culture and Personality Studies," *Trans. N. Y. Academy of Sciences*, ser. II, vol. 7, no. 5 (1945), pp. 128–30.

[24] With P. Oppenheim, "A Definition of 'Degree of Confirmation,'" *Philosophy of Science* 12 (1945), 98–115.

[25] "On the Nature of Mathematical Truth," *The American Mathematical Monthly* 52 (1945), 543–56.

[26] "A Note on the Paradoxes of Confirmation," *Mind* 55 (1946), 79–82.

[27] With P. Oppenheim, "Studies in the Logic of Explanation," *Philosophy of Science* 15 (1948), 135–75.

[28] With P. Oppenheim, "Reply to David L. Miller's Comments," *Philosophy of Science* 15 (1948), 350–2.

[29] "Problems and Changes in the Empiricist Criterion of Meaning," *Revue Internationale de Philosophie* 11 (1950), 41–63.

[30] "A Note on Semantic Realism," *Philosophy of Science* 17 (1950), 169–73.

[31] "The Concept of Cognitive Significance: A Reconsideration," *Procedings of the American Academy of Arts and Sciences*, vol. 80, no. 1 (1951), 61–77.

[32] "General System Theory and the Unity of Science," *Human Biology* 23 (1951), 313–22.

[33] *Fundamentals of Concept Formation in Empirical Science. International Encyclopedia of Unified Science*, vol. II, no. 7 (Chicago: University of Chicago Press, 1952). Spanish edition: *Fundamentos de la formación de conceptos en Ciencia empirica* (Madrid: Alianza Editorial, S.A., 1988).

[34] "Problems of Concept and Theory Formation in the Social Sciences," in American Philosophical Association, Eastern Division, *Science, Language, and Human Rights.*, vol. 1 (Philadelphia: University of Pennsylvania Press, 1952), pp. 65–86. (German translation: "Typologische Methoden in den Sozialwissenschaften," in E. Topitsch, ed., *Logik der Sozialwissenschaften* [Köln und Berlin: Kiepenheuer und Witsch, 4. Auflage 1967].)

[35] "Reflections on Nelson Goodman's *The Structure of Appearance*," *Philosophical Review* 62 (1952), 108–16.

[36] "A Logical Appraisal of Operationism," *Scientific Monthly* 79 (1954), 215–20.

[37] "Meaning," *Encyclopedia Britannica* vol. 15 (Chicago: Encyclopedia Britannica 1956 ed.), p. 133.

[38] "Some Reflections on 'The Case for Determinism,'" in S. Hook, ed., *Determinism and Freedom in the Age of Modern Science* (New York: New York University Press, 1958), pp. 157–63.

[39] "The Theoretician's Dilemma," in H. Feigl, M. Scriven, and G. Maxwell, eds., *Minnesota Studies in the Philosophy of Science*, vol. II (Minneapolis: University of Minnesota Press, 1958), pp. 37–98.

[40] "Empirical Statements and Falsifiability," *Philosophy* 33 (1958), 342–8.

[41] "The Logic of Functional Analysis," in L. Gross, ed., *Symposium on Sociological Theory* (Evanston, IL, and White Plains, NY: Row Peterson & Co., 1959), pp. 271–307. (Italian translation published as a monograph: *La logica dell'analisi funzionale* [Trento: Istituto Superiore di Scienze Sociali, 1967].)

[42] "Science and Human Values," in R. E. Spiller, ed., *Social Control in a Free Society* (Philadelphia: University of Pennsylvania Press, 1960), pp. 39–64.

[43] "Inductive Inconsistencies," *Synthese* 12 (1960), 439–69. Also included in B. H. Kazemier and D. Vuysje, eds., *Logic and Language: Studies Dedicated to Professor Rudolf Carnap on the Occasion of his Seventieth Birthday* (Dordrecht: D. Reidel, 1962).

[44] "Introduction to Problems of Taxonomy," in J. Zubin, ed., *Field Studies in the Mental Disorders* (New York: Grune and Stratton, 1961), pp. 3–23. (Also: contributions to the discussion, on subsequent pages.)

[45] *La formazione dei concetti e delle teorie nella scienza empirica* (Milano: Feltrinelli, 1961). (Contains items 33 and 39 of this bibliography, translated and with an introduction by Alberto Pasquinelli.)

[46] "Meaning," *Encyclopedia Americana*, 1961 edition, vol. 18 (New York: American Corporation), pp. 478–9.

[47] "Deductive-Nomological *vs.* Statistical Explanation," in H. Feigl and G. Maxwell, eds., *Minnesota Studies in the Philosophy of Science*, vol. III. (Minneapolis: The University of Minnesota Press, 1962), pp. 98–169. (Czech translation in K. Berka and L. Tondl, eds., *Teorie modelu a modelování* [Prague: Nakladatelství Svoboda, 1967], pp. 95–172.)

[48] "Explanation in Science and in History," in R. G. Colodny, ed., *Frontiers of Science and Philosophy* (Pittsburgh: University of Pittsburgh Press, 1962), pp. 9–33.

[49] "Rational Action," *Proceedings and Addresses of the American Philosophical Association*, vol. 35 (Yellow Springs, OH: The Antioch Press, 1962), pp. 5–23.

[50] "Carnap, Rudolf," *Colliers Encyclopedia*, 1962 copyright, vol. 5 (New York: Crowell-Collier), pp. 457–8.

[51] "Explanation and Prediction by Covering Laws," in B. Baumrin, ed., *Philosophy of Science. The Delaware Seminar*, vol. 1 (1961–62) (New York: Interscience Publishers, 1963), pp. 107–33.

[52] "Reasons and Covering Laws in Historical Explanation," in S. Hook, ed., *Philosophy and History* (New York: New York University Press, 1963), pp. 143–63.

[53] "Implications of Carnap's Work for the Philosophy of Science," in Paul A. Schilpp, ed., *The Philosophy of Rudolf Carnap* (La Salle, IL: Open Court; and London: Cambridge University Press, 1963), pp. 685–709.

[54] *Aspects of Scientific Explanation and Other Essays in the Philosophy of Science* (New York: The Free Press; and London: Collier-Macmillan, Ltd., 1965). (Japanese translation of title essay published as a monograph [Tokyo: Bai Fu Kan, 1967].)

[55] "Coherence and Morality," *The Journal of Philosophy* 62 (1965), 539–42.

[56] "Comments" (On G. Schlesinger's "Instantiation and Confirmation"), in R. S. Cohen and M. W. Wartofsky, eds., *Boston Studies in the Philosophy of Science*, vol. 2 (New York: Humanities Press, 1965), pp. 19–24.

[57] "Recent Problems of Induction," in R. G. Colodny, ed., *Mind and Cosmos* (Pittsburgh: University of Pittsburgh Press, 1966), pp. 112–34.

[58] *Philosophy of Natural Science* (Englewood Cliffs, NJ: Prentice-Hall, Inc., 1966). (Translations: Japanese, 1967; Italian and Polish, 1968; Swedish, 1969; Portuguese and Dutch, 1970; French, 1972; Spanish, 1973; German, 1974; Chinese, 1986.)

[59] "On Russell's Phenomenological Constructionism," *The Journal of Philosophy* 63 (1966), 668–70.

[60] "Scientific Explanation," in S. Morgenbesser, ed., *Philosophy of Science Today* (New York: Basic Books, 1967), pp. 79–88.

[61] "Confirmation, Qualitative Aspects," in Paul Edwards, *The Encyclopedia of Philosophy*, vol. II (New York: The Macmillan Co. and The Free Press, 1967), pp. 185–7.

[62] "The White Shoe: No Red Herring," *The British Journal for the Philosophy of Science* 18 (1967/68), 239–40.

[63] "Maximal Specificity and Lawlikeness in Probabilistic Explanation," *Philosophy of Science* 35 (1968), 116–33.

[64] "On a Claim by Skyrms Concerning Lawlikeness and Confirmation," *Philosophy of Science* 35 (1968), 274–8.

[65] "Logical Positivism and the Social Sciences," in P. Achinstein and S. F. Barker, eds., *The Legacy of Logical Positivism* (Baltimore: The Johns Hopkins Press, 1969), pp. 163–94.

[66] "Reduction: Ontological and Linguistic Facets," in Sidney Morgenbesser, Patrick Suppes, Morton White, eds., *Philosophy, Science and Method: Essays in Honor of Ernest Nagel.* (New York: St. Martin's Press, 1969), pp. 179–99.

[67] "On the Structure of Scientific Theories," in *The Isenberg Memorial Lecture Series 1965–1966* (East Lansing: Michigan State University Press, 1969), pp. 11–38.

[68] "On the 'Standard Conception' of Scientific Theories," in Michael Radner and Stephen Winokur, eds., *Minnesota Studies in the Philosophy of Science*, vol. IV (Minneapolis: University of Minnesota Press, 1970), pp. 142–63, and some contributions to "Discussion at the Conference on Correspondence Rules," *Minnesota Studies in the Philosophy of Science*, vol. 4, pp. 220–59.

[69] "Formen und Grenzen des wissenschaftlichen Verstehens," *Conceptus VI*, 1–3 (1972), pp. 5–18.

[70] "Rudolf Carnap, Logical Empiricist," *Synthese* 25 (1973), pp. 256–68.

[71] "Science Unlimited?" *The Annals of the Japan Association for the Philosophy of Science*, 4 (1973), pp. 187–202.

[72] "The meaning of Theoretical Terms: A Critique of the Standard Empiricist Construal," in P. Suppes et al., eds., *Logic, Methodology and Philosophy of Science*, IV (Amsterdam: North Holland Publishing Company, 1973), pp. 367–78.

[73] "A Problem in the Empiricist Construal of Theories" (in Hebrew, with English summary) *Iyyun* (A Hebrew Philosophical Quarterly) 23 (1972), pp. 68–81 and 267–8 (published in 1974).

[74] "Formulation and Formalization of Scientific Theories: A Summary-Abstract," in F. Suppe, ed., *Structure of Scientific Theories* (Urbana: University of Illinois Press, 1974), pp. 244–54.

[75] "Carnap, Rudolf," *The New Encyclopedia Britannica*, 15th Edition, vol. 2 (Chicago: Encyclopedia Brittanica, Inc., 1974), pp. 877–8; *Macropedia*, vol. 3, pp. 925–26.

[76] *Grundzüge der Begriffsbildung in der empirischen Wissenschaft* (Dusseldorf: Bertelsmann Universitatsverlag, 1974), 104 pp. (German translation of the author's *Fundamentals of Concept Formation in Empirical Science* [Chicago: University of Chicago Press, 1952], enlarged by an additional chapter, not previously published, on theoretical concepts and theory change.)

[77] "The Old and the New 'Erkenntnis,'" *Erkenntnis* 9 (1975), pp. 1–4.

[78] "Dispositional Explanation and the Covering-Law Model: Response to Laird Addis," in A. C. Michalos and R. S. Cohen, eds., *PSA 1974: Proceedings of the 1974 Biennial Meeting of the Philosophy of Science Association* (Dortrecht: Reidel 1976).

[79] "Die Wissenschaftstheorie des analytischen Empirismus im Lichte zeitgenössischer Kritik," *Kongressberichte des XI. Deutschen Kongresses für Philosophie*, 1975 (Hamburg: Felix Meiner Verlag, 1977), pp. 20–34.

[80] *Aspekte wissenschaftlicher Erklärung.* (Berlin and New York: Walter de Gruyter, 1977). (Revised translation of the title essay of item 54 of this list, with a new section on statistical explanation.)

[81] "Dispositional Explanation," in R. Tuomela, ed., *Dispositions* (Dondrecht and Boston: D. Reidel, 1978), pp. 137–46.

[82] "Selección de una teoría en la ciencia: perspectivas analíticas vs. pragmáticas," in *La filosofía y las revoluciones científicas.* Segundo Coloquio Nacional de Filosofía, (Monterrey, Nuevo Leon, México: Editorial Grijalbo, S.A., 1979), pp. 115–35.

[83] "Scientific Rationality: Analytic vs. Pragmatic Perspectives," in T. S. Geraets, ed., *Rationality To-Day/La Rationalité Aujourd'hui* (Ottawa: The University of Ottawa Press, 1979), pp. 46–58. Also remarks in the discussion, loc. cit., pp. 59–66, passim.

[84] "Der Wiener Kreis-eine persoenliche Perspektive," in H. Berghel, A. Huebner, E. Koehler, eds., *Wittgenstein, the*

Vienna Circle, and Critical Rationalism. Proceedings of the Third International Wittgenstein Symposium, August 1978 (Vienna: Hoelder-Pichler-Tempsky, 1979), pp. 21–6.

[85] "Scientific Rationality: Normative *vs.* Descriptive Construals," in H. Berghel, A. Huebner, E. Koehler, eds., *Wittgenstein, the Vienna Circle, and Critical Rationalism.* Proceedings of the Third International Wittgenstein Symposium, August 1978 (Vienna: Hoelder-Pichler-Tempsky, 1979), pp. 291–301.

[86] "Comments on Goodman's 'Ways of Worldmaking,'" *Synthese* 45 (1980), pp. 193–99.

[87] "Turns in the Evolution of the Problem of Induction," *Synthese* 46 (1981), pp. 389–404.

[88] "Some Recent Controversies Concerning the Methodology of Science," *Journal of Dialectics of Nature* (3, 5 Peking, 1981), pp. 11–20. (In Chinese.)

[89] "Der Wiener Kreis und die Metamorphosen seines Empirismus," in Norbert Leser, ed., *Das geistige Leben Wiens in der Zwischenkriegszeit* (Wien: Oesterreichischer Bundesverlag, 1981), pp. 205–15.

[90] "Analytic-Empiricist and Pragmatist Perspectives on Science," (Chinese translation of a lecture given in Peking). *Kexue Shi Yicong* [Collected Translations on History of Science], 1982, no. 1; pp. 56–63, concluded on p. 47.

[91] "Logical Empiricism: Its Problems and Its Changes," (Chinese translation of a lecture given in Peking) *Xian Dai Wai Guo Zhe Xue Lun Ji* [Contemporary Foreign Philosophy; People's Publishing House], vol. 2 (1982), pp. 69–88.

[92] "Schlick und Neurath: Fundierung *vs.* Kohärenz in der wissenschaftlichen Erkenntnis," in *Grazer philosophische Studien*, vol. 16/17 (1983), pp. 1–18.

[93] "Valuation and Objectivity in Science," in R. S. Cohen and L. Laudan, eds., *Physics, Philosophy and Psychoanalysis: Essays in Honor of Adolf Grunbaum.* (Dordrecht, Boston, Lancaster: D. Reidel Publishing Co., 1983), pp. 73–100.

[94] "Kuhn and Salmon on Rationality and Theory Choice," *The Journal of Philosophy* 80 (1983), pp. 570–2.

[95] "Methodology of Science: Descriptive and Prescriptive Facets," Pamphlet Nr. IAS 814-84 in series of lecture texts

published by The Mortimer and Raymond Sackler Institute of Advanced Studies, Tel Aviv University, 1984 (30 pp.). Revised in [102].

[96]   "Der Januskopf der wissenschaftlichen Methodenlehre," in Peter Wapnewski, ed., *Jahrbuch 1983/84*, Wissenschaftskolleg–Institute for Advanced Study–zu Berlin. Siedler Verlag, 1985, pp. 145–157.

[97]   "Wissenschaft, Induktion und Wahrheit," in a brochure published by Fachbereich Wirtschaftswissenschaft der Freien Universität Berlin, "Verleihung der Würde eines Ehrendoktors der Wirtschaftswissenschaft an Prof. Dr. Phil. Carl G. Hempel (University of Pittsburgh) am 10. Dezember 1984." (Published 1985).

[98]   "Thoughts on the Limitations of Discovery by Computer," in Kenneth F. Schaffner, ed., *Logic of Discovery and Diagnosis in Medicine* (Berkeley: University of California Press, 1985), pp. 115–22.

[99]   "Prova e verità nella ricerca scientifica," *Nuova Civiltà Delle Macchine*, Anno IV, nn. 3/4 (15/16) 1986, pp. 65–71 (Roma). (Translation by Patricia Pincini of Lecture given in June 1986 at Locarno Conference under the title "Evidence and Truth in Scientific Inquiry") English summary, p. 149.

[100]   "Provisoes: A Problem Concerning the Inferential Function of Scientific Theories," *Erkenntis* 28 (1988), pp. 147–64; also in Adolf Grunbaum and Wesley C. Salmon, eds., *The Limitations of Deductivism* (Berkeley: University of California Press, 1988), pp. 19–36.

[101]   "Limits of a Deductive Construal of the Function of Scientific Theories," in Edna Ullmann-Margalit, ed., *Science in Reflection*, The Israel Colloquium, vol. 3. (Dordrecht: Kluwer Academic Publishers, 1988), pp. 1–15.

[102]   "On the Cognitive Status and the Rationale of Scientific Methodology," in *Poetics Today* 9, (1988), 5–27.

[103]   "Las facetas descriptiva y valorativa de la ciencia y la epistemología," *Segundo Simposio Internacional De Filosofía* (1981), vol. I (Mexico, 1988), pp. 25–52. (Enrique Villanueva, compilador).

[104]   *Oltre il Positivismo Logico:* Saggi e Ricordi. A cura di Gianni Rigamonti (Roma: Armando Armando, 1989). Italian trans-

lations of twelve selected essays and of a recorded interview with Richard Nollan, 1982.

[105] "Ernest Nagel"; Memorial Note in the 1989 Year Book of the American Philosophical Society (published in 1990).

[106] "Hans Reichenbach Remembered," *Erkenntnis*, 35 (1991), 5–10.

[107] "Il significato del concetto di verità per la valutazione critica delle teorie scientifiche," *Nuova Civiltà delle Macchine*, VIII, 4 (32), 1990, pp. 7–12. (English text, "The Signification of the Concept of Truth for the Critical Appraisal of Scientific Theories," loc. cit., pp. 109–113. [The second word in the submitted typescript was: "significance.]" Reprinted in William R. Shea and Antonio Spadafora, eds., *Interpreting The World* [Canton, MA: Science History Publications, 1992], pp. 121–9.

[108] "Eino Kaila and Logical Empiricism," *Acta Philosophica Fennica* 52 (1992), 43–52.

[109] "Empiricism in the Vienna Circle and in the Berlin Society for Scientific Philosophy: Recollections and Reflections," in Friendrich Stadler, ed., *Scientific Philosophy: Origins and Developments* (Dordrecht/Boston/London: Kluwer Academic Publlishers, 1993), pp. 1–9.

Index